AIMING FOR AN A
IN A LEVEL
PHYSICS

Mark Jones

HODDER
EDUCATION
AN HACHETTE UK COMPANY

Orders: please contact Bookpoint Ltd, 130 Park Drive, Milton Park, Abingdon, Oxon OX14 4SE. Telephone: (44) 01235 827827. Fax: (44) 01235 400401. Email education@bookpoint.co.uk Lines are open from 9 a.m. to 5 p.m., Monday to Saturday, with a 24-hour message answering service. You can also order through our website: www.hoddereducation.co.uk

ISBN: 978 1 5104 2924 6

© Mark Jones 2018

First published in 2018 by
Hodder Education,
An Hachette UK Company
Carmelite House
50 Victoria Embankment
London EC4Y 0DZ

www.hoddereducation.co.uk

Impression number   10 9 8 7 6 5 4 3 2 1

Year   2022 2021 2020 2019 2018

Typeset by Integra Software Services Pvt. Ltd., Pondicherry, India.

Printed in Spain

A catalogue record for this title is available from the British Library.

MIX
Paper from
responsible sources
FSC
www.fsc.org   FSC™ C104740

Contents

Answers to the activities are online at:
www.hoddereducation.co.uk/AforAPhysics

Getting the most from this book

Aiming for an A is designed to help you master the skills you need to achieve the highest grades. The following features will help you get the most from this book.

Learning objectives

> A summary of the skills that will be covered in the chapter.

✓ Exam tip

Practical advice about how to apply your skills to the exam.

Activity

An opportunity to test your skills with relevant activities.

! Common pitfall

Problem areas where candidates often miss out on marks.

The difference between...

Key concepts differentiated and explained.

Annotated example

Exemplar answers with commentary showing how to achieve top grades.

Worked example

Step-by-step examples to help you master the maths skills needed for top grades.

Take it further

Suggestions for further reading or activities that will stretch your thinking.

You should know

> A summary of key points to take away from the chapter.

About this book

The A-grade student

Only about 9% of physics students achieve an A* grade and 20% a grade A at A-level. To obtain the top grades you must not only respond successfully to questions that assess recall of knowledge and its application in a particular context, but also in the forming of judgements in general contexts. These are higher-order skills for which, according to examiners, students tend to be poorly prepared. This book is a course companion designed to support you from the start to the end of your physics course and develop those higher-order thinking skills that are not covered in standard revision guides and textbooks.

Using this book

This book will develop your ability to apply your mathematical and practical skills to unfamiliar contexts, and to draw together different areas of understanding within a single answer. You will develop the skills and confidence to 'know what to do when you don't know what to do'. Ultimately the aim of this book is to enable you to think for yourself, so that you can successfully overcome unexpected and unfamiliar challenges that you will meet in the A-level course and beyond.

The book begins by developing **quantitative skills** (Chapter 1), with emphasis on evaluating complex expressions, answering multiple-stage calculations, and making judgements and drawing conclusions using quantitative analysis.

Chapter 2 focuses on the **reading skills** needed to develop your subject knowledge, how to engage with material by critical evaluation and how to transfer these skills to examination questions.

Chapter 3 develops your **writing skills** to ensure that you can write clear, concise responses which contain judgements and connections between different areas of the specification. There is a special focus on 6-mark extended-response questions, which assess both the quality of your application of physics and your ability to communicate your reasoning.

Chapter 4 covers **practical skills.** You will learn how to apply fundamental principles of practical physics, including appropriate apparatus and techniques to plan effective investigations. Analysis of data and evaluation of results and procedures are covered, along with the skills of drawing conclusions and suggesting appropriate improvements to experimental designs.

Study skills (Chapter 5) uses ideas from neuroscience to explain the most effective ways to learn and recall physics content for the long term. As well as helping you understand the material better, it also explores methods to help you see synoptic links between different areas of the specification.

Chapter 6 compares the different content and methods of assessment of each exam board. It also gives a brief overview of how each exam board assesses higher-order skills and finishes with some top tips for examination success.

The difference between...

Assessment objective	Weighting	Typical exam question
AO1: Demonstrate knowledge and understanding of scientific ideas, processes, techniques and procedures.	35–40% of an AS paper and 30–35% of an A-level paper	Exam questions of this type test core thinking skills and recall of knowledge. They have command words such as: 'state', 'describe', 'give' and 'name'.
AO2: Apply knowledge and understanding of scientific ideas, processes, techniques and procedures: • in a theoretical context • in a practical context • when handling qualitative data • when handling quantitative data	40–45% of both AS and A-level	Applying knowledge is a higher-order thinking skill, so questions may feature unfamiliar examples. They have command words such as: 'explain', 'calculate', 'derive' and 'show'.
AO3: Analyse, interpret and evaluate scientific information, ideas and evidence, including in relation to: • making judgements and reaching conclusions • developing and refining practical design and procedures	25–30% of an AS paper and 25–30% of an A-level paper	Questions may require you to make judgements, reach conclusions and develop and refine practical procedures. They have command words such as: 'assess', 'analyse', 'decide', 'evaluate', 'criticise', 'comment on' and 'discuss'.

How to tackle unfamiliar questions: use MAPS to navigate the question

Often students find it difficult to know how to start a physics question. This is because they don't understand what they are being asked to do. This book is full of practical strategies that you can apply to help you think about and solve problems in unfamiliar contexts.

The following **mnemonic** is a very useful way to analyse questions to the identify the key information that will enable you to write a top-quality answer:

Use MAPS to navigate the question:

M	Marks	How many marks are allocated? This gives you information about how much detail is required or whether a calculation will involve multiple stages or the use of appropriate significant figures etc.
A	Action	What action does the command word tell you to take? 'Describe' requires a very different response from 'explain' which is very different from 'discuss'. The action tells you what type of response is required.
P	Physics principles	What are the underlying physics principles in the question? What particular equations or ideas are involved?
S	Situation	What is the context of the question? How does the physics concept apply in this specific situation?

Combining this strategy with effective reading skills, as discussed in Chapter 2, will enable you to quickly and efficiently identify what type of response is expected. The information in Chapter 3 will help you be able to compose a clear, coherent response which is appropriate to the question.

The key to developing your ability to apply your knowledge to the deep structure of the question is to practise critically evaluating every question you tackle. When revising don't just analyse the question you have been asked, but also think about how the question could have been asked differently using the information presented, or think about how the question could be extended, either to assess a different concept or to increase the level of difficulty of the question — what other information would you need to be given? Practising this skill will give you invaluable experience in applying and evaluating questions and will improve your understanding of physics, enabling you to see the whole and not just the parts.

1 Quantitative skills

Introduction

Quantitative skills require the use of numerical data in a variety of formats. Developing the skills to manipulate and analyse data in order to draw conclusions is essential if you are to be successful on the A-level course and beyond. At least 40% of the assessment in your A-level (and AS) physics course will require the use of mathematical skills equivalent to Level 2. Level 2 is a grade 9–4 on the new GCSE grading system (or A*–C on the older grading system).

You should aim to develop mastery of the core mathematical skills required on the course, which means that you should be able to apply them automatically. Examples are recalling scientific prefixes, rearranging equations and use of appropriate significant figures.

This quantitative chapter will begin by introducing the mathematical skills required on the course. The first activity allows you to try out many of the required skills without the distraction of a contextualised question. The chapter then discusses the core quantitative skills in more detail. The core mathematical skills are those which can be learned prior to the examination. There are worked examples to show you how to demonstrate your knowledge of the appropriate skills.

The chapter then goes on to look at higher-order study skills. Application style questions are characterised by extended calculations and applying your understanding to explain quantitative changes. To reach an A/A* grade you need to master the ability to use quantitative analysis to evaluate new situations and make reasoned judgements. The development of these skills is the focus of the final part of the chapter.

Core study skills
Knowledge and understanding

Ten core mathematical skills

Questions asking you to demonstrate your knowledge and understanding of mathematical skills require you to recall information and skills that can be learned prior to the examination. The following are ten core skills which you must be able to carry out automatically and with minimum effort:

1 Recall appropriate units for all physical quantities and be able to convert between units, e.g. mm → m, cm² → m², radians → degrees.
2 Convert between standard form and ordinary form.
3 Calculate a mean.
4 Use estimates to check calculated values.
5 Use a calculator for power, exponential, logarithmic and trigonometric functions.
6 Rearrange and solve algebraic equations.
7 Substitute numerical values into equations (using appropriate units).
8 Plot two variables on a graph, draw a line of best fit, calculate the gradient of the line and determine the intercept.
9 Give answers to an appropriate number of significant figures.
10 Calculate circumference, surface area and volume of regular shapes.

> ✅ **Exam tip**
>
> Physical quantities consist of a value matched to a unit, so if there is no unit beside the answer space, include one.

> ✅ **Exam tip**
>
> **Prefixes**
>
> The following prefixes and their symbols must be learned:
> - pico, p ×10^{-12}
> - nano, n ×10^{-9}
> - micro, μ ×10^{-6}
> - milli, m ×10^{-3}
> - centi, c ×10^{-2}
> - deci, d ×10^{-1}
> - kilo, k ×10^{3}
> - mega, M ×10^{6}
> - giga, G ×10^{9}
> - tera, T ×10^{12}

> ✅ **Exam tip**
>
> Most equations are given on the data and formulae sheet, but three that you do need to recall are:
>
> - specific charge $= \dfrac{\text{charge}}{\text{mass}}$
> - volume of a cylinder = cross-sectional area × length
>   Since the diameter, d, is the most commonly measured quantity of a wire or string, the most useful form of this equation is often:
> $$V = \frac{\pi d^2 l}{4}$$
> - volume of rectangular block = length × width × height
> - stopping potential, $V_S = \dfrac{E_{kmax}}{e}$

Four quick tips that apply to all quantitative questions

1 **Where possible, use the symbols that are given to you in the question.** For example, M for mass of a planet, or g_s for the gravitational field strength at the planet's surface.
2 **Show your substitution of values into the equation.**
3 **Substitute values into the equation before rearranging.** This gives the examiners a chance to give you a mark for this process and also generally results in fewer mistakes in the rearrangement.
4 **Set out your working clearly.** This enables the examiner to give marks for stages in your working even if your final answer is incorrect. It also enables you to identify and correct mistakes more effectively in your calculations.

Activity 1.2

You should be able to recall the correct units for all quantities in the equations on the data sheet. Occasionally you may not be able to remember the exact unit — for example, what is the unit for resistivity?

When this happens, go back to the equation and ensure that the quantity you are looking for is the subject of the equation:

$$R = \frac{\rho l}{A}$$

$$\rho = \frac{RA}{l}$$

The units must match on both sides of the equation:

$$= \frac{\Omega\, m^2}{m}$$

Simplifying, we get:

$$= \Omega\, m$$

Thus, the unit for resistivity, ρ, is the ohm metre, $\Omega\, m$.

Try these questions:

1 Determine the units of

$$\frac{2\varepsilon - V_A}{V_B}$$

where ε is the emf of the cell and V_A and V_B are the potential differences across components A and B.

2 Determine the units of $\sqrt{l + x} - \sqrt{l}$, where l is the length of a string and x is the extension.

3 Determine the units of the Young modulus.

Answers online – see page 3

Worked example 1.1

This example combines recall of basic information relating to atomic structure, such as atomic notation and specific charge, with the mathematical skills of standard form, significant figures and ratios.

An isotope of oxygen can be represented as $^{18}_{8}O$.

(a) Calculate the number of neutrons. (1)

(b) Calculate the specific charge of the nucleus. Give your answer to an appropriate number of significant figures. (3)

(c) The ratio of the specific charge of the nucleus of this isotope to the specific charge of the nucleus of a heavier isotope of oxygen is 1.4:1. Determine the number of neutrons in this heavier isotope of oxygen. (3)

(a) **Step 1:** Marks, Action, Physics Situation (MAPS):

   M: 1 — this is a very simple calculation.

   A: 'calculate' means we need to manipulate numbers to generate a numerical answer.

✅ Exam tip

SI units

The Système International (SI) base quantities and their units are:
- mass in kilograms, kg
- length in metres, m (and therefore area in m^2 and volume in m^3)
- time in seconds, s
- current in amperes, A
- temperature in kelvin, K
- amount of substance in moles, mol

✅ Exam tip

You can use dimensional analysis to check that the form of an equation is physically sensible or to check the units of an unknown quantity.

The fundamental units used for dimensional analysis are:
- length [L]
- time [T]
- mass [M]
- charge [C]
- temperature [Θ]

P: you must recall that the top number is the mass of the isotope, which is is the number of protons and neutrons. The bottom number is the number of protons (also called the atomic number). The number of neutrons can be calculated as the difference between the mass number (protons and neutrons) and the number of protons.

S: the mass number is 18 and the atomic number is 8.

Step 2: Carry out the calculation:

mass number – atomic number = 18 – 8 = 10 neutrons ✓

(b) Step 1: M: 3 — more than just a simple calculation, so you need to consider what else is required in the question (see below).

A: 'calculate' means you need to substitute values into an equation to produce a numerical answer.

P: you need to recall the equation to calculate the specific charge of any nucleus or ion:

$$\text{specific charge} = \frac{\text{charge}}{\text{mass}}$$

S: the question is asking for the specific charge of a nucleus, so you do not need to consider the number of electrons. Therefore the charge comes only from the 8 protons and the mass comes only from the protons and neutrons (not the electrons). Because the mass of the proton and the neutron are the same to three significant figures (1.67×10^{-27} kg) we consider both types of particle to have this mass.

The question asks for the final answer to be written to an **appropriate number of significant figures** (this is the extra mark to make it a 3 mark question instead of a more straightforward 2 mark calculation).

Step 2: Carry out the calculation:

$$\text{specific charge} = \frac{\text{charge}}{\text{mass}}$$
$$= \frac{8 \times 1.60 \times 10^{-19}}{18 \times 1.67 \times 10^{-27}} ✓$$
$$= 42\,581\,503\,C\,kg^{-1} ✓$$

There are two points to make about the value we have calculated:
- Large and small numbers should always be written in standard form.
- All final answers should be written to an appropriate number of significant figures.

Although you should keep the full figure either on your page or in your calculator memory for use in subsequent calculations, you need to format your final answer correctly. First you need to decide the appropriate number of significant figures — the rule is to use the same as the lowest number of significant figures in the question. The working uses data from the data sheet and both values are given to three significant figures. Therefore your answer also needs to be written to three significant figures:

> ✓ **Exam tip**
>
> Become familiar with the content on the data sheet by using it when you answer questions throughout the course — this will save you time in the exams.

$$\text{specific charge} = 42\,600\,000\,\text{C}\,\text{kg}^{-1}$$

Although technically this is the correct answer and it is written to the correct number of significant figures, you should write this large number in standard form:

$$\text{specific charge} = 4.26 \times 10^7 \, \text{C}\,\text{kg}^{-1} \checkmark$$

(c) **Step 1:** **Marks, Action, Physics, Situation (MAPS):**

M: 3 — multiple stage calculation.

A: 'Determine' — this is a calculation, but not a straightforward substitution of values into equations.

P: specific charge $= \dfrac{\text{charge}}{\text{mass}}$, so the heavier isotope will have a lower specific charge.

S: the specific change of the lighter isotope calculated in part (b) is 1.4 times greater than the specific charge of the heavier isotope, but the charge is the same so you can work out the mass in kilograms and then divide by the mass of a nucleon to find the total mass.

Step 2: Calculate specific charge of heavier isotope:

$$\text{specific charge of new isotope} = \frac{4.26 \times 10^7}{1.4}$$
$$= 3.04 \times 10^7 \, \text{C}\,\text{kg}^{-1} \checkmark$$

Step 3: Calculate mass of new isotope:

$$\text{mass} = \frac{\text{charge}}{\text{specific charge}}$$
$$= \frac{8 \times 1.6 \times 10^{-19}}{3.04 \times 10^7}$$
$$= 4.21 \times 10^{-26} \, \text{kg}$$

Step 4: Determine number of nucleons:

$$\text{number of nucleons} = \frac{\text{mass}}{\text{mass per nucleon}}$$
$$= \frac{4.21 \times 10^{-26}}{1.67 \times 10^{-27}}$$
$$= 25.1 = 25 \text{ nucleons} \checkmark$$

Step 5: Determine number of neutrons in this isotope of oxygen:

$$\text{number of neutrons} = \text{mass number} - \text{atomic number}$$
$$= 25 - 8$$
$$= 17 \text{ neutrons} \checkmark$$

Worked example 1.2

You are expected to be able to use percentages in any context, so here is a different example in the form of a percentage change.

(a) Determine the mass per unit length of an oscillating string if the density is 1340 kg m⁻³ and it has a uniform diameter of 4.0×10^{-4} m. **(2)**

> ✓ **Exam tip**
>
> In any quantitative question you should check that your final answer seems sensible. Go back and check your working if not. Clear presentation of your calculations will make it far easier to locate the problem in your working.

> ✓ **Exam tip**
>
> Always show your full working out. Calculations in the multiple-choice questions will require you to do working (although you do not gain extra marks for this).

(b) Suppose the wire, which is initially 2000 mm in length, is put under tension which results in a 0.42% increase in length. Calculate the increase in volume if there is no appreciable change in the diameter of the string. (2)

(a) Step 1: Marks, Action, Physics, Situation (MAPS):

M: 2.

A: 'Determine' — this is asking for a quantitative line of reasoning.

P: The concept of volume is often combined with density. Density is the mass per unit volume of a substance:

$$\rho = \frac{m}{V}$$

Unit: kg m^{-3}

This equation is often most useful in the form of $m = \rho V$. This is often used when calculating the mass of a planet (assuming the density of the planet is constant).

The equation $m = \rho V$ is also useful when calculating the mass per unit length, μ, of a material such as a string:

If $\mu = \frac{m}{l}$ and since $m = \rho V$:

$$\frac{m}{l} = \frac{\rho V}{l}$$

$\mu = \rho A$ This is another useful relationship to remember.

S: you have been given the (uniform) diameter, d, of the wire, so you can calculate the area, A, either by $A = \frac{\pi d^2}{4}$ or $A = \pi r^2$ (where r = radius = $\frac{d}{2}$).

Step 2: Substitute the values into the equation:

$$\mu = \rho A = \rho \, \pi \left(\frac{d^2}{4}\right) = 1340 \times 3.14 \times \frac{(4.0 \times 10^{-4})^2}{4} \checkmark$$

Step 3: Calculate the final value:

$$\mu = 1.68 \times 10^{-4} \, \text{kg m}^{-1} \checkmark$$

(b) Step 1: Marks, Action, Physics, Situation (MAPS):

M: 2.

A: 'calculate' — use an appropriate equation to find a numerical answer.

P: an increase in a quantity of 0.x% means that the new length is 100.x% not x% longer. A wire is a cylinder and the volume of a cylinder can be calculated using: $V = A \times l$. The change in volume of a cylinder with constant diameter can be calculated using: $\Delta V = A \times \Delta l$.

S: there is a '0.42% increase in length' means that the length is now 100.42%. This is *not* a 42% increase in length. The new length is 1.0042 × 2000 mm = 2008.4 mm = 2008 mm because the length measurement must have been made with a long measuring device such as a tape measure, which has a resolution of ±1 mm (see Chapter 4 for more on this).

Exam tip

In a multiple-part question you should always use your full answer in the subsequent stages of the question. The exception to this is if the earlier part was a 'show that' question and your answer is nowhere near the 'show that' value. In that case you should use the 'show that' value in the further calculations.

Step 2: Substitute the values into the equation for change in volume.

$$\Delta V = A \times \Delta l$$

$$\Delta V = \pi \left(\frac{d^2}{4}\right) \times \Delta l$$

$$\Delta V = 3.14 \times \frac{(4.0 \times 10^{-4})^2}{4} \times 8 \times 10^{-3} \; \checkmark$$

Step 3: Calculate the change in volume:

$$\Delta V = 1.0 \times 10^{-9} \, \text{m}^3 \; \checkmark$$

Higher-order study skills

Application of knowledge

The skills we have looked at so far involve recalling rules or stating units, describing simple procedures, and knowing equations (e.g. for gradient). Questions that assess higher-order skills in physics are characterised by requiring you to contextualise your knowledge and understanding. This means that your responses are not a rewriting of your revision notes, but you need to use your understanding to explain what is happening, or what might happen in a specific **situation**. Application of knowledge questions account for the largest percentage of the marks across your A-level (42%).

Each examination question has a specific action or command word which tells you what type of response is required. The key **action** words for AO2 quantitative questions are:

→ **Calculate** Use an appropriate equation to find a numerical answer.

→ **Determine** This is asking for a quantitative line of reasoning.

→ **Derive** Combine at least two equations and rearrange to find an algebraic expression.

→ **Explain** Describe what is happening or what the effect would be and then give the reasons behind your statement. Qualitative 'explain' questions often involve the application of a rule such as Fleming's left-hand rule.

→ **Show** Often these questions involve a numerical calculation — you must write your answer to at least one extra significant figure than the 'show that' number in the question.

You will need to routinely demonstrate the following skills:

→ Select the appropriate equation from the data sheet.

→ Substitute values into the equation.

→ Ensure the substituted values are in standard units, e.g. $570 \, \text{nm} = 570 \times 10^{-9} \, \text{m}$.

→ Calculate the final value.

→ Select an appropriate number of significant figures for your final answer.

→ Write your final value with the appropriate unit.

These stages often form the majority of student practice questions during class and homework activities. One problem is that often students use books that only quote the final answer. This puts an emphasis only on the final answer and promotes poor-quality working, both in terms of the detail and the clarity and flow of the working.

A/A* grade-level questions will often involve extended, multiple-stage calculations.

Worked example 1.3

This question relates to the photoelectric effect.

When monochromatic ultraviolet radiation of wavelength 210 nm is incident upon a metal, electrons are emitted with a maximum kinetic energy of 8.7×10^{-19} J. Calculate the work function of the metal and explain what will happen to the velocity of the emitted electrons if the frequency of the incident radiation is doubled. (5)

Step 1: Marks, Action, Physics, Situation (MAPS):

M: 5 — but these marks are distributed across both aspects of the question.

A: 'calculate' is the command word for the first part and the second part requires you to 'explain' what happens when part of the system is changed.

P: the photoelectricity equation is $hf = \phi + E_{kmax}$. Each photon interacts with only one electron. If the energy of the incoming photons is greater than the work function, then the number of emitted electrons is proportional to the intensity.

S: you are given the wavelength rather than the frequency, so you can either use $c = f\lambda$ to find the frequency and use the photoelectric equation, or replace $E = hf$ with $E = \dfrac{hc}{\lambda}$. Typical photoelectric effect questions ask you to explain the relationship between incident photon frequency and number of electrons emitted per second. However, here you are asked to explain the effect on the speed of the emitted electrons. Recall that the prefix nm is nanometres: $\times 10^{-9}$.

Step 2: Select the appropriate equation and substitute the values (use the data sheet for the numerical value of the constants h and c):

$$hf = \phi + E_{kmax}$$

Substitute hf with $\dfrac{hc}{\lambda}$:

$$\frac{hc}{\lambda} = \phi + E_{kmax}$$

$$\frac{6.63 \times 10^{-34} \times 3 \times 10^8}{210 \times 10^{-9}} = \phi + 8.7 \times 10^{-19} \checkmark$$

Step 3: Rearrange to calculate the work function, ϕ:

$$\phi = \frac{6.63 \times 10^{-34} \times 3 \times 10^8}{210 \times 10^{-9}} - 8.7 \times 10^{-19}$$

$$= 9.47 \times 10^{-19} - 8.7 \times 10^{-19}$$

It can be useful to write down the calculations completed at each stage so that the examiner can clearly see your working. If you make a mistake the first steps are often worth marks.

$$\phi = 7.71 \times 10^{-20} \text{ J } \checkmark$$

Step 4: Write your final answer to an appropriate number of significant figures. In this case all data in the question are given to two significant figures, so this should also be applied to the final answer.

$$\phi = 7.7 \times 10^{-20} \text{ J}$$

Step 5: Look at the next part of the question. What happens to the velocity if the frequency is doubled?

Doubling the frequency doubles the energy of the incident photons. $\checkmark$

This will double the maximum value of the kinetic energy of the emitted electron (because the work function will remain the same), i.e. the kinetic energy will be: 1.74×10^{-18} J. $\checkmark$

$E_k = \frac{1}{2}mv^2$

Rearrange to make v the subject: $v = \frac{2E_k}{m}$, so the velocity of the emitted electron will increase by a factor of $\sqrt{2}$, i.e. 1.41. $\checkmark$

In this case it is important to use values where possible, i.e. saying that the kinetic energy doubles rather than just saying it increases (and definitely don't just say it changes). This allows you to determine exact relationships without having to make all of the calculations — i.e. here you do not need to calculate the kinetic energy of the emitted electrons with both frequencies of light to find the exact way that the velocity will change.

Take it further

Use online assessment resources such as Isaac Physics https://isaacphysics.org/ to improve your quantitative problem-solving skills. The immediate feedback you receive on your incorrect answers will help you learn from your mistakes more rapidly.

Analysis, interpretation and evaluation

Each examination question follows a staircase or ramping model whereby, as the question progresses, the difficulty increases. Generally the question begins with knowledge domain questions (AO1) worth 1 or 2 marks, such as asking you to state a definition or complete a simple calculation. The question then begins to assess application of knowledge by increasing the level of difficulty of the calculations (AO2). The final part of the question is the most demanding and requires you to make a judgement and reach conclusions (AO3). To reach appropriate conclusions and judgements you will need to analyse, interpret and evaluate the information you have developed in your answers and new information supplied at this stage of the question. Students often find these kinds of question the toughest on the exam paper.

Let's look first at the meaning behind these key higher-order action/command words:

→ **'Assess'** Consider all factors that might apply and then decide which are the most important or relevant. You are making a judgement on relative importance and then drawing a conclusion.

→ **'Analyse'** This means separate information into components and identify their characteristics. You need to also be able to discuss pros and cons of the idea or device and make a reasoned comment.

→ **'Decide'** You must provide a statement as to your decision along with the evidence that you have used to make that decision.

→ **'Evaluate'** Review all available evidence and make a judgement. You will need to also use your own knowledge and understanding in order to help you identify strengths and weaknesses in the particular context being examined.

→ **'Interpret'** This means that you have to 'translate information into a recognisable form'. Most commonly this means that you will have to explain the meaning of information presented in graphical form (e.g. a stress–strain graph), or in the form of a diagram (e.g. energy levels in an atom).

Other action/command words assessing higher-order skills in quantitative questions are:

→ **'Deduce'** This means 'draw conclusions from the information provided'. This may involve different skills from simply calculating — e.g. you may be expected to obtain information from a graph and apply this to an equation.

→ **'Criticise'** Here you have to inspect numerical data, statements, scientific ideas or experimental methods and make a judgement about their strengths and weaknesses.

→ **'Comment on'** This means that you have to synthesise a variety of information and form a judgement.

→ **'Discuss'** You should address a range of ideas and arguments. In terms of quantitative questions you do not need to calculate exact values, but you should explore the effects on different factors in terms of increases or decreases.

→ **'Compare and contrast'** Identify at least one similarity and one difference between two (or more) things. You do not need to come to an overall conclusion. Remember to address both (or all) things stated in the question.

> ✓ **Exam tip**
>
> Action/command words give you an indication of the style of response you should give to the question. The rest of the wording in the question gives the context of the response and may also contain other key words. Remember to read all of the words in the question and to consider the number of marks allocated before beginning your response.

Worked example 1.4

Derive an expression for the escape velocity of an object of mass _m_ from a planet of mass _M_ and radius _R_. Comment on how your result could explain why some planets have an atmosphere and others do not. (4)

This question has two parts. First we need to derive an expression for escape velocity. Escape velocity is the minimum velocity given to an object at the planet's surface that will allow it to leave the planet's gravitational field. It is important to understand that this velocity is needed if there is no further input of energy, but if the object can be supplied with energy — for example by burning rocket fuel — then the object can escape the gravitational field without reaching escape velocity.

The derivation is an application of conservation of energy. When an object escapes the gravitational field we say it is at infinity and the gravitational potential energy has its maximum value of zero. The gravitational potential energy decreases as you get closer to the planet so that it has a value of $E_p = \dfrac{-GmM}{R}$ at the planet's surface.

The initial kinetic energy at the surface must be at least equal to this initial amount of gravitational potential energy. ✓

$\frac{1}{2}mv^2 = \dfrac{GmM}{R}$ ✓

We can cancel the _m_s and rearrange this equation to find the minimum value of velocity:

$v = \sqrt{\dfrac{2GM}{R}}$

If $v > \sqrt{\dfrac{2GM}{R}}$, then the particle will have enough velocity to escape ✓. Planets of larger mass have larger escape velocities and so will be more likely to retain an atmosphere. Planets of smaller mass have a lower value for the escape velocity, so gas particles are more likely to exceed the escape velocity, leaving the planet without an atmosphere ✓.

 Exam tip

Always use the symbols given to you in the question when deriving expressions.

Worked example 1.5

Comparing an unfamiliar equation given to you in an exam to the equation for a straight line ($y = mx + c$) is one of the most common methods for processing data to calculate other quantities.

This example looks at a graphical method of analysis to determine the internal resistance of a cell.

You are given a series circuit with a cell, filament lamp, variable resistor, ammeter and voltmeter.

(a) Explain how you could use experimental data to determine the emf, _ε_, and the internal resistance, _r_, of a cell. (5)

(b) Discuss the effect of replacing the original lamp with a lamp of twice the resistance. (3)

(c) Explain the consequences of adding an identical cell to the original circuit:

   (i) in series

   (ii) in parallel   **(3)**

(a) From the data sheet, the emf can be calculated using:

$$\varepsilon = I(R + r)$$

where R is the resistance of an external resistor in the circuit and r is the internal resistance of the circuit.

$$\varepsilon = V + Ir$$

We can measure V, the potential difference across the lamp, using the voltmeter connected in parallel with the lamp. We can measure the current, I, in the series circuit using the ammeter. ✓

Use variable resistor to obtain a range of values for p.d. and current. ✓

Rearranging the equation to make V the subject, we get:

$$V = \varepsilon - Ir$$

If we compare this to the equation for a straight line we get:

$$V = -rI + \varepsilon$$

$$y = mx + c$$

where m is the gradient of the line and c is the y-intercept.

If we plot a graph of V against I ✓ (i.e. V on the y-axis and I on the x-axis), we will get a straight-line graph with a negative gradient i.e. slopes from top left to bottom right.

ε is the intercept on the y-axis. ✓

r is minus the gradient of the graph. ✓

The second part of the question is asking us to discuss the effects of doubling the resistance of the lamp. This means that we need to explore the implications on various factors, but we do not need to calculate exact values.

If the lamp is replaced with a lamp of twice the resistance, the current in the circuit will reduce ✓:

$$I = \frac{\varepsilon}{R + r}$$

The potential difference across the internal resistance will decrease: Ir. ✓

A smaller proportion of the energy will be wasted by the internal resistance, so the overall efficiency of the circuit will increase. ✓

The final part of the question requires the rules about combining cells in series and parallel:

(i) When the two cells are combined in series the emfs of the two cells add together: $\varepsilon_T = \varepsilon_1 + \varepsilon_2$ ✓ and the internal resistances also add up, just like the equation for resistors in series $r_T = r_1 + r_2$ ✓.

(ii) When the extra cell is added in parallel to the first the total emf, ε, is unaffected. However, the internal resistance will halve $\left(\frac{1}{r_T} = \frac{1}{r_1} + \frac{1}{r_2}\right)$ decrease so the current will double. ✓

Worked example 1.6

Logarithms are an excellent tool for evaluating complex expressions where you have quantities raised to powers. The following example uses periodic motion.

The following expression relates a particular oscillating system's time period, T, to its length, l.

$$\frac{1}{T^2} = kl^m$$

where k is a constant and m is an integer.

Evaluate this expression to determine the value of m and k. (3)

To see relationships between variables in an equation we need to use linear graphs. When there are unknown powers involved — i.e. m in our equation — we need to use logarithms.

Step 1: Take logs of both sides:

$$\log\left(\frac{1}{T^2}\right) = \log(kl^m)$$

$$= \log k + \log l^m$$

$$= \log k + m\log l$$

Step 2: Compare this expression to the equation for a straight line:

$$y \quad = mx \quad + c$$

$$\log\left(\frac{1}{T^2}\right) = m\log l + \log k \checkmark$$

So if you plot a graph of $\log\left(\frac{1}{T^2}\right)$ against $\log l$, you should get a straight line of positive gradient. The gradient of the graph gives the value of m in the expression. $\checkmark$

You would need to include the negative sign if the graph had a negative gradient. Often the value will be an integer as in our question, so you will need to round your gradient value to the nearest whole number.

Step 3: Identify the y-intercept with k:

$\log k$ represents the y-intercept of the line of best fit. This gives us an equation:

$\log k = y$-intercept

Step 4: Use the reverse of the log function to get k on its own. This is called the antilog function.

You can see where it arises from using the following examples:
→ $\log 1000 = 3$ and (using antilog) $10^3 = 1000$
→ $\log 100 = 2$ and $10^2 = 100$

So if: $\log k = y$-intercept,

$10^{(y\text{-intercept})} = k \checkmark$

Activity 1.3

$$f = \frac{1}{2l}\sqrt{\frac{T}{\mu}}$$

What would the gradient of a $\ln f$ against $\ln \mu$ graph be?

Answers online – see page 3

Application to the exam

The following examination style question brings together the ideas covered in this quantitative chapter. You must use your quantitative analysis skills to evaluate complex expressions, work through multistage calculations and use your quantitative results, combined with your understanding of the underlying physics principles, to give insight into unfamiliar situations.

A satellite of mass 1330 kg and cross-sectional area 1.5 m^2 is in a low polar orbit at a height of 198 km above the Earth's surface.

(a) **Show that the satellite makes 16 orbits for each full rotation of the Earth.** **(5)**

(b) **The density, ρ, of the atmosphere can be modelled in a simple way using an exponential equation:**

$$\rho = \rho_0 e^{-h/a}$$

where

ρ_0 **is the density of air at sea level, 1.23 kg m^{-3}**

h **is the height above the Earth's surface**

a **is the effective height of the atmosphere if it were to have a constant density with altitude 8.42 km**

Use the equation to estimate the density of the atmosphere at the orbit height of the satellite. **(2)**

(c) **According to NASA the drag force, F_d, can be considered to be the main force on low-Earth-orbiting (LEO) spacecraft. The drag force can be calculated using:**

$$F_d = \tfrac{1}{2}\rho A C_d v^2$$

where A = cross-sectional area of the satellite and C_d = drag coefficient = 2.2.

Calculate the drag force on the satellite. **(2)**

(d) **Calculate the work done against friction per 24 hours.** **(2)**

(e) **Use the information from the question and graph in Figure 1.1 to evaluate the issues with low-Earth-orbiting (LEO) satellites.** **(3)**

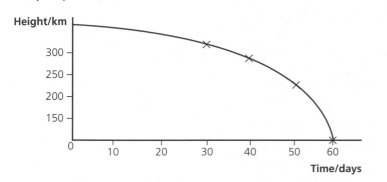

Figure 1.1 Altitude graph for an LEO satellite

How to approach the question

(a) The first part of the question requires us to find the time period or velocity of the satellite and show that it allows 16 complete orbits in 24 hours.

(b) Use the exponential density equation given in the question to calculate the density at a height of 198 km above the Earth's surface.

(c) To use the drag equation ($F_d = \frac{1}{2}\rho ACdv^2$) we first need to calculate the speed of the satellite.

This can be done several ways, but the easiest is probably just to use $v = \frac{d}{t}$.

Then substitute values into the drag equation.

(d) Identify the appropriate equation. Work done can be calculated using the equation for work on the data sheet:

$W = Fs\cos\theta$

If we consider the force to always be acting in the opposite direction to the velocity, this reduces to:

$W = Fs$

where s is the distance travelled in 24 hours.
Substitute values into the equation, $W = Fs$.

(e) Describe the pattern shown in the graph.

Explain why this process occurs.

Evaluate LEO orbits compared with other orbits.

B-grade answer

(a) $F = \frac{mv^2}{r} = \frac{GMm}{r^2}$

$v = \sqrt{\frac{GM}{r}}$

$v = \sqrt{\dfrac{6.67 \times 10^{-11} \times 5.97 \times 10^{24}}{6.37 \times 10^6}}$

$v = 7906\,\text{m s}^{-1}$ ✓ ecf

distance in one revolution = circumference of circle, $C = 2\tfrac{1}{2}r$

$C = 2 \times 3.14 \times 6.37 \times 10^6 = 4 \times 10^7$

$d = s \times t = 7906 \times 24 \times 60 \times 60 = 6.8 \times 108$ ✓ ecf

revolutions = $d/C = 17$ ✓ ecf Therefore the satellite can make at least 16 complete revolutions.

This candidate scores 3 marks. The processes they have carried out are correct, but they have failed to take into account the radius of the Earth in their distance. This has resulted in them not arriving at the correct 'show that' figure. They should also display their final answer to at least one extra significant figure in a 'show that' question.

(b) $\rho = 1.23 \times e^{-198/8.42}$ ✓

$\rho = 7.53 \times 10^{-11}\,\text{Pa}$ ✓

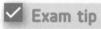

> ✓ **Exam tip**
>
> In a 'show that' question make your working out clear and always show your final numerical answer to at least one more significant figure than the value given in the question.

Both marks awarded. There is even less chance of making a mistake if the original equation is written out before substituting the values.

(c) $F_d = 0.5 \times 7.53 \times 10^{-11} \times 1.5 \times 2.2 \times (7.9 \times 10^3)^2$ ✓ *ecf*

$F_d = 9.81 \times 10^{-7}\,N$ ✗

They have correctly substituted values into the equation (they receive 1 mark and error carried forward because they have used their incorrectly calculated velocity from part (a). However, in the subsequent calculation they have forgotten to square the velocity.

(d) work = force × distance = $9.81 \times 10^{-7} \times 6.8 \times 10^8$ ✓ *ecf* = $667\,J$ ✓ *ecf*

Both marks awarded, although both values in the equation are actually incorrect due to previous errors in their calculations.

(e) The graph shows a quick descent of the satellite (in less than 60 days), because the density of the atmosphere increases. This means that very low-Earth-orbiting satellites will crash towards Earth quickly.

No marks awarded here. None of the points in the mark scheme is addressed with sufficient clarity or detail. In the first point it looks like the candidate thinks the satellite has crashed after 60 days, but since the y-axis does not start at the origin this is not necessarily true. The candidate mentions that the density of the atmosphere increases, but does not use information from earlier parts of the question to do this accurately enough (i.e. they do not mention the exponential nature of this increase in pressure). The candidate tries to make a comment about the issues with LEO satellites, but it is too vague.

A/A*-grade answer

(a) Centripetal force on the satellite is caused by the gravitational force of the Earth:

$F = m\omega^2 r = \dfrac{GMm}{r^2}$ ✓

$\omega^2 = \dfrac{GM}{r^3}$

$r = r_E + h = 6.37 \times 10^6 + 1.98 \times 10^5 = 6.57 \times 10^6\,m$ ✓

$\omega^2 = \dfrac{6.67 \times 10^{-11} \times 5.97 \times 10^{24}}{(6.57 \times 10^6)^3}$

$\omega^2 = 1.41 \times 10^{-6}$

$\omega = 1.19 \times 10^{-3}\,rad\,s^{-1}$ ✓

since $\omega = \dfrac{2\pi}{T}$, then $T = \dfrac{2\pi}{\omega}$

$T = \dfrac{2 \times 3.14}{1.19 \times 10^3}$

$T = 5.30 \times 10^3\,s$ ✓

In 24 hours there are $60 \times 60 \times 24 = 86\,400\,s$

The number of orbits in 24 hours = $86\,400/5.30 \times 10^3 = 16.3$ ✓ orbits, so it can make 16 complete orbits (but not 17).

There are many steps in this calculation and it is easy to make mistakes at various stages. The clear way in which the working has been done not only makes it easy for the examiner to award marks, but the student can easily identify where the error is if the final answer does not yield the 'show that' value.

(b) $\rho = \rho_0 e^{-h/a}$

$\rho = 1.23 \times e^{-198/8.42}$ ✓

$\rho = 7.53 \times 10^{-11}$ Pa ✓

Full and clear working shown. Both marks awarded.

(c) $v = \dfrac{d}{t} = \dfrac{(16.3 \times 2 \times 3.14 \times 6.57 \times 10^6)}{86\,400}$

$v = 7783\,\text{m s}^{-1}$

$F_d = \dfrac{1}{2}\rho A C_d v^2$

$F_d = 0.5 \times 7.53 \times 10^{-11} \times 1.5 \times 2.2 \times (7.78 \times 10^3)^2$ ✓

$F_d = 7.5 \times 10^{-3}\,\text{N}$ ✓

The working is clearly shown in this multistage calculation, and the full values are used for latter parts of the calculation.

(d) $W = Fs$

$W = 7.5 \times 10^{-3} \times (16.3 \times 2 \times 3.14 \times 6.57 \times 10^6)$ ✓

$W = 5.1 \times 10^6\,\text{J}$ ✓

Both marks awarded. Again, clear working and accurate use of the calculator.

(e) The graph shows that for a satellite the rate of descent increases as the satellite descends (this indicates that descent is exponential). ✓ This is because the satellite is doing a greater amount of work against the atmosphere at lower altitudes because the density increases exponentially as the altitude decreases. ✓ LEO satellites will always have more drag than those at higher orbits so they will 'lose energy' by doing work on the surroundings and lose altitude more quickly than at higher Earth orbits, which means they will need to be replaced more frequently. ✓

A very good response, for all 3 marks. The candidate correctly describes the altitude change over time from the graph and then explains why this behaviour occurs, using their understanding of physics and the ideas from the earlier parts of the question. They finish their response by evaluating the issues relating to this behaviour with regard to real-life, low-orbit satellites.

You should know

> There are many core mathematical skills that you are expected to use throughout A-level Physics questions. Ensure you can carry out these processes routinely.

> Always set out your working clearly and logically. This not only makes mistakes less likely, but it also allows you to find errors more quickly and allows examiners to award marks for stages in your work, especially in longer multiple-stage calculations.

> You can use dimensional analysis to check that the form of an unfamiliar equation is physically sensible or check the units of an unknown quantity.

> Logarithms are a useful tool for evaluating expressions with quantities raised to powers.

> Comparing complex expressions to the equation of a straight line allows effective analysis of the different parts of the expression.

> Use the action word in the question to decide on the type of mathematical analysis you carry out.

> Higher-level questions require you to evaluate your final numerical answer and make judgements not only about the implications of this data, but also about how valid the arguments in the analysis are, i.e. what assumptions have been made?

2 Reading skills

Learning objectives

> To develop effective note taking techniques
> To research effectively and cite sources
> To apply appropriate reading skills to different situations
> To use critical reading to evaluate sources of information
> To apply effective reading skills to examination questions

Introduction

This chapter aims to help you develop a range of reading skills which will not only improve your performance in exams but also give you the skills necessary to learn effectively. The step up from GCSE, and subsequent steps into degree and further study or employment, require a greater degree of independent studying. This is reliant on effective reading: recalling, processing and evaluating information.

The chapter begins by looking at the core resources you should be using to support your studies and then examines how different reading styles should be used to maximise your comprehension of the text. You will then be guided through the very important skill of taking effective notes and referencing your sources when carrying out research for your practical endorsement. The chapter then explores the higher-order thinking skills of evaluating and engaging with material through critical reading and explores how wider reading can push you towards an A/A* grade.

Core study skills

Using core resources

It is important to make effective use of the resources available to you. The key resources you need initially are the **specification** for your examination board, the **data and formulae sheet** and the **practical handbook**. These important documents are available to download from the website of your examination board.

The specification is the definitive guide to the content you will be examined upon. The headings from the specification make a logical way of organising your notes and also help to ensure that you have covered every required part of the specification. You should work with your data and formulae sheet whenever you are answering questions — familiarity with the location of content on the equation sheet as well as what quantities the symbols represent will

Activity 2.1

Go to your examination board's website and download a copy of the specification. In addition to defining the examinable material, the specification also gives information about the content and length of exams, and the data and formulae sheet. The specification will also enable you to track your progress through the course and identify areas you would like your teacher to help with. You may find it useful to print off relevant pages to keep in your A-level file/exercise book.

save you valuable time in exams. The practical handbook is the definitive guide to practical techniques and processing data. There is more detail on this resource in Chapter 4 of this book.

Most A-level students will also use a textbook to extend and support work done in class. If you use a textbook that covers multiple boards, then carefully check the sections that apply to your specification.

Most textbooks contain 'test yourself' questions for each section. The answers are usually supplied at the back of the book. Make sure that you attempt these questions for every section of the course that you have covered. These types of question will help you develop mastery of the equations and basic principles of each topic. However, remember that their emphasis will be on AO2, focusing on calculations, with some AO1 content involving recall of definitions and simple statements.

Effective reading

Being able to learn from the information provided in texts is not about reading quickly, but it is about reading effectively. To read effectively you must do the following:

→ read with purpose — have a **clear goal** for the reading
→ choose the right **reading material**
→ select the right **reading style**
→ use appropriate **note taking** techniques to ensure longer-term recall

Read with purpose

First and foremost you need to determine *why you are reading*. By determining a reading goal, you make it easier to identify the relevant material in the text you are reading and allow yourself to spend more time on the key information, thus maximising your reading effectiveness.

Choose the right reading material

Once you have identified why you are reading, you then need to use suitable reading material to find the required information. The most important thing is that the source is reliable — i.e. the information provided is factually correct. Material is considered reliable if it is from a reputable peer-reviewed source, or by a well-known publisher. There is more information on critically evaluating sources for reliability in the higher-order study skills section.

Reading styles

You use different styles of reading on a daily basis. For example, reading information from your social media feed requires a different reading style from reading questions in an end-of-topic assessment. The main three types of reading style are:

→ scanning
→ skimming
→ intensive reading
→ extensive reading

> **! Common pitfall**
>
> Often students copy out notes from the textbook or classwork in full. This is ineffective in many ways: it takes longer to write out everything; you are not processing the information; (by summarising and reorganising) so you will not retain it; and it does not promote links with your prior knowledge. Use the strategies and tips in this book to make your note taking more effective.

Scanning

Scanning is a technique that is useful when you are trying to locate specific information within a text. Your eyes pass rapidly over the text to try to locate particular words or phrases.

For example, if you were looking for a different point of view on gravitational potential, you would scan the contents page to find the chapter on gravitational fields, then you would scan through the chapter to find the description, equation and examples related to gravitational potential. You can use other contextual information, such as key words being written in a bold typeface or the headings and subheadings of the text, to help you scan more effectively.

Skimming

Skimming is where you read quickly for general meaning. Skimming is used to get an understanding of the main points, while skipping over words containing detailed information.

It is often useful to skim a page of text or a section of a textbook before reading it in more detail. Once you have an overview of the material, it should be easier to see how the specific information you encounter in your more detailed reading fits within the 'bigger picture' and this will also help to structure your note taking. You can also use skimming to refresh your understanding after you have completed a more detailed reading.

Intensive reading

Intensive reading is where you examine the content in detail with a specific goal in mind. You may be revising the content for a test or exam or you might be researching a topic that you will be covering in a subsequent lesson. To be effective, intensive reading should be active rather than passive.

Active reading

Simply reading and rereading is not an effective way to understand or remember information. In essence, active reading means interacting and engaging with the material in different ways so that you critically evaluate your understanding of the material and see how the information connects to your prior knowledge.

Try these techniques to make your reading more active:

→ Underline or highlight key words and phrases as you read. Be selective though — a full page of highlighted material will not make key information stand out.

→ Make a note of key information, questions or ideas as you read — you could annotate in the margins or use sticky notes.

→ Summarise the key information — use bullet points, mind maps, or flashcards to keep your notes condensed.

→ Test yourself by putting the text away and making a note of the key points from memory. This is useful for identifying gaps in your understanding.

→ Explain what you have learned to someone else, e.g. a family member or study partner.

> ✓ **Exam tip**
>
> When reading intensively underline key information, e.g. when given an angle for a force or a light ray, make sure you have noted whether it has been given to the horizontal or to the vertical.

→ Read the information critically — this is a great way of engaging with the text and making you think more deeply about the content (more guidance on this is contained in the 'Critical reading' section).

Activity 2.2

Visit your school library and choose one of the 'popular' physics books — you can ask the librarian for help locating these. Read the book and make some brief notes about what you found interesting. Can you see any links to the work you are doing in class?

Extensive reading

Extensive reading is reading texts for pleasure. The goal of this kind of reading is enjoyment, so you do not need to make notes or write questions as you read. You should choose material that is interesting and engaging, such as magazine articles or popular science books — your school library should have a selection of these and you can speak to your physics teacher to get other recommendations.

Extensive reading enables you to develop your scientific vocabulary, build your background knowledge and develop your own ideas and opinions. It may even help you uncover areas which you may wish to study further, e.g. cosmology or particular areas of engineering.

Note taking

Every student needs to take notes at some point in order to process and retain information. There are lots of ways to try to remember the content of your course. **The key to remembering information is thinking about the information.** This means that you have to think about your notes as you make them, i.e. don't just copy information from the board, textbook or website. If you process the information, by rearranging it, summarising it and writing it in your own words, you will not only have a better understanding of the content, but you will remember it for longer.

Your notes should be considered a 'work in progress' and can always be updated. Key phrases from the mark schemes and examiner's reports as you take past paper questions make an extremely useful addition to your initial set of notes. You may find that when you reread your notes from earlier in the course you will be able to reduce the content and summarise further.

You should organise your notes in a logical order. As suggested above, the key headings from the specification make a good set of topic titles to structure your notes.

✓ Exam tip

Definitions, equations and units should be learned as they are presented by your teacher/ textbook and not re-written in your own words.

Annotated example 2.1

The following example shows how information presented on one page of a textbook (Figure 2.1) can be processed to make effective notes. Three key strategies are involved: summarising, re-structuring, and using key questions to check content and detail.

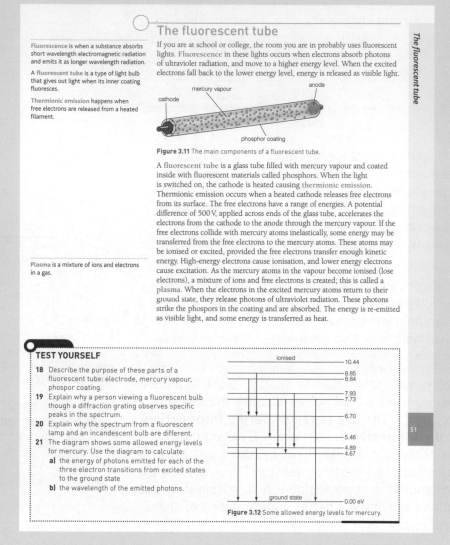

The fluorescent tube

Fluorescence is when a substance absorbs short wavelength electromagnetic radiation and emits it as longer wavelength radiation.

A fluorescent tube is a type of light bulb that gives out light when its inner coating fluoresces.

Thermionic emission happens when free electrons are released from a heated filament.

If you are at school or college, the room you are in probably uses fluorescent lights. Fluorescence in these lights occurs when electrons absorb photons of ultraviolet radiation, and move to a higher energy level. When the excited electrons fall back to the lower energy level, energy is released as visible light.

Figure 3.11 The main components of a fluorescent tube.

A fluorescent tube is a glass tube filled with mercury vapour and coated inside with fluorescent materials called phosphors. When the light is switched on, the cathode is heated causing thermionic emission. Thermionic emission occurs when a heated cathode releases free electrons from its surface. The free electrons have a range of energies. A potential difference of 500 V, applied across ends of the glass tube, accelerates the electrons from the cathode to the anode through the mercury vapour. If the free electrons collide with mercury atoms inelastically, some energy may be transferred from the free electrons to the mercury atoms. These atoms may be ionised or excited, provided the free electrons transfer enough kinetic energy. High-energy electrons cause ionisation, and lower energy electrons cause excitation. As the mercury atoms in the vapour become ionised (lose electrons), a mixture of ions and free electrons is created; this is called a plasma. When the electrons in the excited mercury atoms return to their ground state, they release photons of ultraviolet radiation. These photons strike the phospors in the coating and are absorbed. The energy is re-emitted as visible light, and some energy is transferred as heat.

Plasma is a mixture of ions and electrons in a gas.

TEST YOURSELF

18 Describe the purpose of these parts of a fluorescent tube: electrode, mercury vapour, phospor coating.
19 Explain why a person viewing a fluorescent bulb though a diffraction grating observes specific peaks in the spectrum.
20 Explain why the spectrum from a fluorescent lamp and an incandescent bulb are different.
21 The diagram shows some allowed energy levels for mercury. Use the diagram to calculate:
    a) the energy of photons emitted for each of the three electron transitions from excited states to the ground state
    b) the wavelength of the emitted photons.

Figure 3.12 Some allowed energy levels for mercury.

51

Figure 2.1 Sample page from AQA A-level Physics student textbook

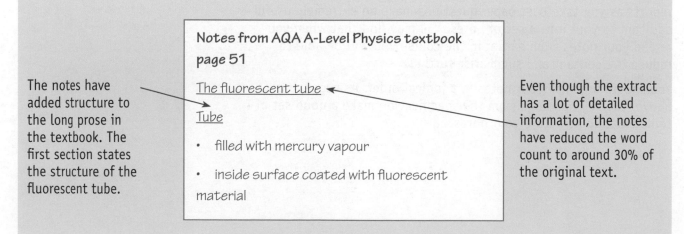

Notes from AQA A-Level Physics textbook page 51

The fluorescent tube

Tube

• filled with mercury vapour

• inside surface coated with fluorescent material

The notes have added structure to the long prose in the textbook. The first section states the structure of the fluorescent tube.

Even though the extract has a lot of detailed information, the notes have reduced the word count to around 30% of the original text.

Exciting mercury atoms

Free electrons from cathode accelerated by high voltage

They collide with electrons in mercury atoms, transferring energy to them and causing:

- excitation (lower energy free electrons)

  or

- ionisation (high energy)

Creates plasma of mercury ions and free electrons

Emitting light

When electrons in excited mercury atoms return to ground state, release UV photons

UV (short wavelength EM rad) photons hit phosphor coating and move electrons to higher energy level

Excited electrons fall back to lower energy level → emit visible light (and some energy transferred as heat)

Comparison with incandescent bulb

Fluorescent tube has emission line spectra because only certain transitions allowed — between discrete energy levels in the mercury atoms

Incandescent bulb emits continuous spectrum because heated filament has continuous range of energies

This section explains the process of exciting the mercury atoms by electron–electron collisions.

This section explains how the UV photons emitted from the mercury atoms are converted into visible frequency photons.

The notes have a clear, logical structure, which would enable them to be applied directly in an extended-response exam question.

The final section of the notes uses the questions in the 'Test yourself' section to check content coverage (question 18) and add further detail. The process of comparing the output from the fluorescent tube with an incandescent bulb (questions 19 and 20) ensures a deeper exploration of the material and will result in a longer-term retention of the information. Making comparisons and finding links to other areas of the specification also allows you to try and anticipate the type of higher-level questions you will face in the exam.

Researching new sources of information

There are many reasons why you would want to find new sources of information about your A-level content. Perhaps you have found it difficult to understand a certain topic from notes and discussion in class — you have looked in your textbook, but you are still unsure. In this case a different point of view would be helpful. Another reason is that in order to achieve a pass in the Practical Endorsement, you will need to have conducted some research. Criterion 5b of the Common Practical Assessment Criteria (CPAC) states that you need to 'cite sources of information demonstrating that research has taken place, supporting planning and conclusions'.

Take it further

Read up on the 'faster than light neutrinos' experiment. In September 2011, researchers published results showing muon neutrinos travelling faster than the speed of light, a result which violated Einstein's special theory of relativity. Reading up on this experiment will give you insight into how the physics community reacts to controversial new findings and will also serve as a good review of particle physics.

Referencing your sources

Referencing is an important skill because it allows you to acknowledge the words and ideas from another author in your work. When you progress to university, referencing is mandatory. Not only does it give credit to scientists who carried out earlier research, but it also allows others reading your work to evaluate the breadth and depth of the prior research you have carried out on the subject you are writing about.

You need to reference when:

→ you **quote directly** from a source
→ you **paraphrase** ideas (i.e. you are writing the ideas in your own words)

Although it can be useful to record the source of information when researching for revision notes so that you can find the original source, the most important time for referencing on the the A-level Physics course is when carrying out research for your required practical activities. Most often this will involve paraphrasing, rather than quoting text directly. Either you will be describing different experimental methods that you can use to carry out a practical activity or you will be using results or values obtained by a different experimenter.

How to reference sources

There are two types of referencing, when paraphrasing and when quoting exact phrases. There are many styles of referencing, but the mostly widely used is the Harvard style, which will be explained below:

1   Paraphrasing

When you are paraphrasing a source you just need to indicate the author and the year of publication, e.g. (Jones, 2018).

2   Exact quotes

When you are quoting text exactly, you need to add quotation marks around the text and add the page number to the reference. For example: 'Wider reading has educational benefit but it also should be an enjoyable activity, free from the pressures of deadlines and assessments.' (Jones, 2018, p35)

Whenever you cite research in your work, you must also include the full source. This can either be done at the bottom of the page or in a separate section at the end called 'references' or 'bibliography'.

For a printed or electronic book this is set out as follows:

Surname, Initials. (Year of publication) *Title in Italics*, Edition if not first, Place of publication: Publisher

Example:

Jones, M. (2018) *Aiming for an A in A-level Physics*. London: Hodder Education.

For a webpage:

Surname, Initials [or organisation responsible for the site]. (Year created or last updated), Name of sponsor of site (if available), accessed day month year (the date you viewed the site), URL or internet address (between pointed brackets)

Example:

Hodder & Stoughton. About Us, accessed 20 December 2017 <https://www.hodder.co.uk/Information/About%20Us.page>

Worked example 2.1

In your lab book you need to demonstrate your ability to cite sources of information and demonstrate that you have conducted research to support your planning.

You have been asked to research a method to investigate acceleration due to gravity using a free-falling object.

Step 1: Find reliable sources of information.

There are many places you could go to get a method, but it is important that the information you find is reliable. Material is considered reliable if it is from a reputable peer-reviewed source or by a well-known publisher.

Suppose we choose the OCR Practical Assessment book by Hodder Education. The extract in Figure 2.2 gives a suggested method and diagram of the apparatus:

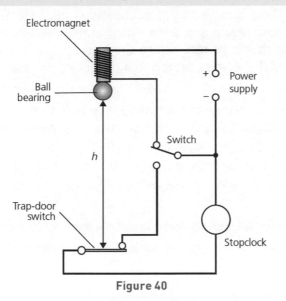

Figure 40

Figure 40 shows one possible method. The points to remember are as follows:

- The time is measured using an electronic stopwatch that reads to 1/100th of a second.
- When the switch is in the position shown, the electromagnet is activated and holds the steel ball bearing in place.
- When the switch is thrown, the electromagnet circuit is broken, releasing the ball, and simultaneously the clock circuit is activated, starting the clock.
- The clock is stopped when the ball hits the trapdoor and breaks the circuit.
- The time should be determined at least three times and averaged for each height, over as wide a range of heights as possible.

Figure 2.2

Step 2: **Read the material and extract the relevant information to generate a suitable a plan for the experiment.**

Method:
1 could use the following apparatus to measure acceleration due to gravity (Davenport, C. and George, G., 2017):

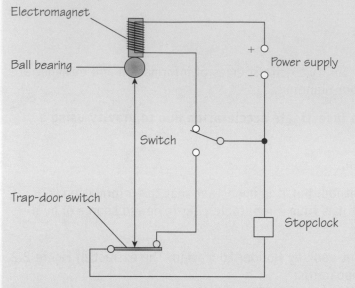

You need to be specific with your method. There is more on this in Chapter 4, but basically you must state the range of readings, the interval of your readings and the instruments used to take the readings. Also remember that any experimental values should be recorded to the resolution of the measuring device. For example, length using a metre ruler should be stated to a resolution of 1 mm (0.001 m).

1   Start with a distance, h, of 1.000 m between the electromagnet and the trap-door switch.
2   The switch will initially be in the closed position. This completes the top circuit and current flows through the electromagnet causing it to be activated and attract the steel ball bearing.
3   Press the switch to the lower position, causing the current in the electromagnet to stop. This causes the ball bearing to be released and also starts the stopclock. When the ball bearing falls into the trap-door switch the bottom circuit is broken. This stops the stopclock.
4   Record the fall time in a results table.
5   Repeat the experiment two more times for this height and calculate a mean value for the time.
6   Reduce the height by 0.100 m and carry out the timing experiment three more times.
7   Repeat this for a range of values from 1.000 m to 0.200 m.

Then you should also describe how you will process these measurements.
Plot a graph of t^2 against h.

The gradient of the graph should be equal to $\frac{2}{g}$, where g is the acceleration due to gravity.

The intercept should be zero, but if it is not zero, there must be a systematic error in my results.

Either, the t values are too big because there was a delay in releasing the ball even though the timer had started, or a delay in the trap door opening when it is hit by the ball bearing.

Or, the values of h are too small so there was a systematic error in measuring the height, h.

You would then add the book or resource into your references section at the end of the lab report.

References
Davenport, C. and George, G. (2017) *OCR Physics A and B Practical Assessment*, London: Hodder Education.

Higher-order study skills

Reading beyond the specification

Students who want to be successful at A-level and make a strong application to a top university tend to have explored physics through wider reading outside the classroom. This sort of reading is often called 'super-curricular' because it enhances the knowledge and understanding you have from studying in school. Remember that you do not need to purchase any books. There are excellent free web resources available and it is also a good idea to visit your school or local library to see the resources they have on offer. You can discuss your particular area of interest or a particular author with the librarian and they can suggest relevant materials or order books that you would like to look at.

The great thing about 'super-curricular' reading is that you can follow your own passion and take the time to explore areas of the subject you are most interested in. Not only will this give you a greater understanding of what you might like to study beyond your A-levels, but it will also help you to build connections between the material you are covering on your A-level course and its wider application.

There are many other ways to explore physics beyond the specification. You could visit science museums, or attend lectures, seminars or other outreach activities from local universities. You might join a regional or national club or society — many areas have amateur astronomy societies. There are many factual television and radio programmes available via BBC iPlayer. There are excellent science podcasts on the internet, and many universities offer audio recordings of their academic lectures. Use a podcast app on your mobile device or search for podcasts on university websites.

Wider reading has educational benefit, but it also should be an enjoyable activity, free from the pressures of deadlines and assessments.

Take it further

Project Tuva: Richard Feynman's Messenger Lecture series has been described by Bill Gates as the 'best science lectures I've ever seen'. Explore the Feynman lectures at:
http://research.microsoft.com/apps/tools/tuva/index.html

Critical reading

Through your A-level studies and beyond, you are expected to develop a critical and analytical mind, so that you do not simply take on information at 'face value'. You need to apply the technique of critical reading when looking at new information. Critical reading is not about being negative or closed off about an argument, but about engaging with, and making judgements about, what you are reading.

The following sections give a range of questions you should consider when critically reading a text:

Author

→ Who are the writers?

→ What are their credentials?

→ Are they scientists? If not, have they consulted scientists with sufficient qualifications? What institutions are they affiliated with?

→ Why have they written it? Who is funding the research and who may profit from it?

→ Is the work cited by others?

Source

→ Who owns or edits the website?

→ Is the same story reported in different ways in different sources?

→ Is the source a **peer-reviewed** scientific journal?

→ Is the source a published book? If so the research will have been reviewed by the publisher.

Hypotheses

→ Are there testable and falsifiable hypotheses? Scientific theories should allow predictions to be made and evidence to be collected to test the predictions. Do they describe what results would be expected if the hypothesis was false?

Procedure and results

→ What steps were taken to control the other variables?

→ Were repeat measurements taken?

→ Are the results reproducible?

→ Are there any results which might undermine the findings?

Conclusions

→ How valid are the conclusions? (With what degree of certainty and in what circumstances are the conclusions valid?)

→ Could you draw different conclusions from the evidence?

→ What other studies could be done to be more certain about the conclusions?

→ Are there clear lines of reasoning in the explanation or does the author try to use scare tactics and other persuasive techniques?

Reading exam questions

Many exam questions contain a lot of introductory information to process before you can answer the questions effectively. Use the reading skills described in this chapter to help you gain understanding of the new introductory material presented in the question and of what is being asked of you in the high-pressure exam situation.

Activity 2.3

Research the risks and benefits of nuclear power. Explore a range of different websites, ensuring that you find some that are in favour of and some against nuclear power. Read your sources critically using the guidance in the 'Critical reading' section of this chapter. Use your research to evaluate the balance between the risks and benefits of the development of nuclear power.

1 Read with purpose

You need to have a clear understanding of what you are being asked to do and use any additional information in the text to tailor your answer to the specific question you are being asked.

The following mnemonic is an excellent way to remember why you are reading the material:

Use MAPS to navigate the question:

M — Marks — how many marks are allocated?

A — Action — what action does the command word tell you to take?

P — Physics — what are the underlying physics principles of the question?

S — Situation — what is the context of the question: how does the physics concept apply in this situation?

2 Select the right reading style

Step 1: **Skim** the stem text and subsequent questions to get an overview of the **physics principles (P)** involved.

Step 2: Read the question stem **intensively.** Here you should highlight or underline key words and add relevant notes to the question paper, e.g. mark missing information on to the diagram.

Step 3: **Scan** the question part you are about to answer to identify the **command word**. This indicates the **action (A)** that you need to take to produce a response in line with the marking points expected by the examiner.

Step 4: **Scan** the question to determine the number of **allocated marks (A)** for this part of the question. The number of marks, rather than the amount of space allocated, is the best guide for the quantity of information required in your answer.

Step 5: Read the question **intensively** to ensure you fully understand what is being asked. One of the most common comments from examiners is that students do not answer the specific question being asked. Respond to the particular **situation (S)** being referred to in the question.

> ☑ **Exam tip**
>
> **'Calculate by making suitable measurements on the Figure...'** (5)
> There are two key parts to understand from the wording in such a question:
> * The command word is 'calculate', so we need to process numerical data to obtain a numerical answer.
> * The next part of the introduction to the question states 'making suitable measurements on the diagram'. Many candidates lose marks by not clearly showing on the diagram the measurements that they need to make.

Worked example 2.2

Careful reading of the not statement

Often multiple-choice questions will ask which of the following is **not** correct, or in which of the following situations does a certain phenomena not occur. Although the key word, not, is written in a bold typeface, many students still jump for the first correct answer they see.

In which of the following situations does electromagnetic induction not take place? (1)

A   in a hydroelectric power station

B   in an ac power adaptor for a mobile phone

C   in a back-up capacitor for a computer

D   in the wings of an aircraft as it cuts through the Earth's magnetic field lines

Many students will rush in and select the first option. An easier option has deliberately been placed at the beginning so that you are drawn into thinking that it is an easy question and you move quickly on to the next question. If you take a little more time to read the second option, you will then see that electromagnetic induction also takes place in the step-down transformer of the power adaptor. This would then make you analyse the last two options in more detail and perhaps reread the question to catch that it asks for the option where electromagnetic induction does *not* take place. C deals with only dc electricity, so electromagnetic induction does not take place here. Many students may still go for D because it is not obviously to do with electromagnetic induction: until we realise that Faraday's law of electromagnetic induction states that the value of the emf induced depends upon the rate of change of flux and how many loops of conductor are in the field.

Application to the exam

To remain in a geostationary orbit a satellite needs to be 35 800 km above the Earth's surface. Currently, it costs about £5000 per kilogram to send objects to this height using a rocket propulsion system. The two major contributors to the cost are the large amount of propellant and the fact that the rocket is not reusable. The propellant makes up 80% of the launch mass. The propellant contains fuel, but is mostly made of oxidiser, which is liquid oxygen. A further 16% of the launch mass is the rocket itself, which leaves 4% available for cargo, i.e. experimental apparatus, astronauts, food etc.

Two new technologies have been proposed to get objects into space at a greatly reduced financial cost: space planes and a space gun.

Space planes would have a higher efficiency because they use oxygen from the Earth's atmosphere as the oxidiser rather than carrying liquid oxygen with them.

A space gun does not need to take any propellant. The energy required for launch is provided by an explosion within a long cannon. The spacecraft would reach a speed of $17\,000\,\mathrm{m\,s^{-1}}$ in 10 seconds.

(a) Calculate the minimum energy required to send a payload of 800 kg into a geostationary orbit. Ignore the mass of the propellant. **(3)**

(b) Acceleration due to gravity on the Earth's surface, $g = 9.81\,\mathrm{m\,s^{-2}}$. On a roller coaster you may experience $5g$ and a fighter pilot can typically handle around $9g$, but a constant $16g$ for anything other than a brief moment may be deadly to humans. Determine whether the space gun is viable for human space flight. **(3)**

(c) Discuss the role that space planes and space guns could have in making space travel more affordable. **(4)**

How to approach the question

There are several parts to this question, all involving processing information from the stem of the question.

Step 1: **Skim** through the question to identify the underlying **physics (P)** principles — there is a lot of new information presented here, so it helps to get an overview first. It is a wide-ranging question involving energy changes: chemical or kinetic to gravitational potential energy.

Step 2: **Intensively** read the information to pick out the key details for each question.

Step 3: **Scan** the question to identify the **action (A)** word:

Part (a) is '**calculate**', so we need to use an equation and obtain a numerical answer.

Part (b) is '**determine**', so we need to use a numerical argument and come to a conclusion.

Part (c) is '**discuss**', so we need to address the benefits and drawbacks of the new technologies introduced in the question. We need to intensively read the section of text again, focusing on the issues impacting on the cost of spaceflight. There are several key information points to pick up on: costs can be reduced by using less propellant and reusing the rocket. We then need to talk about both new technologies in comparison with the current rocket launch process.

Mark scheme

(a) payload = 800 kg = 4%

mass of rocket plus pay load = 20%

therefore, 800 × 5 = 4000 kg ✓

$\Delta W = m\Delta V$

$\left[= mGM\left(\frac{-1}{r} - \frac{-1}{R_E}\right)\right]$

$\Delta W = GMm\left(\frac{1}{R_E} - \frac{1}{r}\right)$ ✓

$= 6.67 \times 10^{-11} \times 5.97 \times 10^{24} \times 4000$

$\times \left(\frac{1}{6.37} \times 10^6 - \frac{1}{(6.37 \times 10^6) + (35\,800 \times 10^3 ✓)}\right)$

$= 2.12 \times 10^{11}$ J ✓

(b) Acceleration can be calculated using: $a = \frac{v-u}{t} = \frac{17\,000 - 0}{10}$

$= 1700\,\text{m s}^{-2}$ ✓

Determine the number of gs: $\frac{1700}{9.81} = 173g$ ✓

This amount of g-force is above the threshold for human survival, so the space gun is not a suitable option for human spaceflight. ✓

(c) Relevant points relating to propellant and/or reusability and cost implications for both space planes and space guns.

Indicative content:

Space planes:
- Less propellant needed so efficiency savings.
- More cargo or passengers can be taken into space.
- Fewer trips to carry cargo would need to be taken.
- Even a 1% reduction in fuel mass, means that you could have an extra 1% total mass for cargo. In real terms this means that you would have an extra 25% cargo.
- Launching from a higher altitude could mean that there is less stress on the rocket and it may be more reusable.

Space guns:
- Require much less fuel (because you have a greatly reduced mass of spacecraft).
- The overall mass of the spacecraft would be reduced by around 80%.
- However, humans cannot be launched into space using this technology — the high g-forces would be lethal to living creatures. However, you could space guns to launch non-living materials, such as building supplies and foodstuffs (to then be manipulated and used by humans who used the current space launch technology).
- This could reduce overall costs.
- Not clear that the spacecraft would be reusable after such an explosive launch. **(Max 4)**

Sample answer

(a) Potential, V at Earth's surface:

$$V_s = \frac{-GM}{R_E}$$

$$= \frac{-6.67 \times 10^{-11} \times 5.97 \times 10^{24}}{6.37 \times 10^6}$$

$$= -6.25 \times 10^7 \, J\,kg^{-1}$$

Potential, V, of geostationary orbit:

$$V_g = \frac{-GM}{r}$$

$$= \frac{-6.67 \times 10^{-11} \times 5.97 \times 10^{24}}{(6.37 \times 10^6) + (35\,800 \times 10^3)} \checkmark$$

$$= -9.44 \times 10^6 \, J\,kg^{-1}$$

$$\Delta V = (-9.44 \times 10^6) - (-6.25 \times 10^7)$$

$$= 5.31 \times 10^7 \, J\,kg^{-1} \checkmark$$

total energy change $= \Delta V \times$ mass of payload

$$= 5.31 \times 10^7 \times 4000$$

$$= 2.12 \times 10^{11} \, J \checkmark$$

This candidate has broken the calculation down into stages. This is a good approach to minimise errors, but it does make for a slightly longer calculation. They have correctly calculated the geostationary orbital radius by adding the radius of the Earth to the height above the surface.

(b) Use SUVATs to calculate acceleration:

$s = X$

$u = 0$

$v = 17\,000\,m\,s^{-1}$

$a = ?$

$t = 10\,s$

$v = u + at$

$$a = \frac{v - u}{t}$$

$$= \frac{17000 - 0}{10}$$

$$= 1700\,m\,s^{-2} \checkmark$$

$$\frac{a}{g} = \frac{1700}{9.81} = 173.3 \checkmark$$

Since a g-force of 16 could be fatal, the g-force exerted on passengers from the space gun would be very likely to cause death and therefore it would not be safe for human flight. $\checkmark$

The candidate has correctly calculated the acceleration, although they have used a longer method than simply using $a = \Delta v/\Delta t$, and they have correctly calculated the g-force. A sensible conclusion relating to the information given in the question satisfies the final marking point.

(c) Both technologies will save money for space flight. They both use less propellant (because they need to take less oxygen) and therefore they are not as heavy and will therefore not cost as much to launch. However the space gun is not suitable for passengers so could only fire supplies and parts into space, whereas the space plane could transport humans and supplies. ✓

The answer is not well structured and is not specific enough in the points it addresses. Each alternative should be discussed separately and in more detail. One marking point is achieved for identifying that the space gun could be used for launching supplies, but not living objects.

> ✓ **Exam tip**
>
> **Multiple-choice technique**
>
> In questions that require a statement to be chosen, think carefully, and read through *all* of the statements carefully. Don't rush to a conclusion before reading them all.

You should know

> > Your exam board specification and core textbook are useful resources for independent study. Use these to track your progress and support classwork. Make use of the summary questions and exam-style questions in the textbook.

> > When you read new information, do so critically. This means that you analyse and evaluate the information and the source and do not just accept everything that you read as true.

> > Enjoy your 'super-curricular' reading. There are lots of books, websites and magazines that will excite and engage you as well as increasing your background knowledge of physics.

> > When reading an exam question, identify the command word and use the information given to you in the question to produce an effective answer.

3 Writing skills

Learning objectives

> To write clear, concise responses which demonstrate knowledge and understanding
> To write using precise, technical language
> To present information using clear, unambiguous language
> To communicate ideas and information effectively using an appropriate style of writing
> To give coherent, structured answers in extended-response questions
> To make connections between different areas of physics to answer synoptic questions
> To develop your ability to make judgements using analysis and evaluation

✓ **Exam tip**

Remember that 40% of the A-level course overlaps with the GCSE Physics course. This means that if a question looks straightforward, then it often is. You need to answer such questions efficiently, which means quickly and concisely. Look carefully at the number of marks available and then do not overcomplicate your answer. Efficient responses to simpler questions will allow you to spend more time on the more challenging parts of the paper.

Introduction

Communication is a vital life skill, whether it is with friends and family, at a job interview or in your university application. The aim of this chapter is to help you develop the skills necessary to communicate your knowledge and understanding effectively to an examiner.

We will look at selecting the appropriate style to communicate your ideas and the use of appropriate terminology. We will also look at how to respond to questions that require an extended response. Extended-response questions require you to demonstrate your ability to construct and develop a sustained line of reasoning which is coherent, relevant, substantiated and logically structured. As part of your extended response you may also be required to include extended calculations. We go through the key skills for writing effective longer responses.

In addition, we will look at the higher-order skills of making judgements, and answering synoptic questions which tackle different areas of the curriculum in one question. The chapter finishes with an example question along with different student responses for you to see how the skills in the chapter apply in examination situations.

! **Common pitfall**

When questions have a lot of information in them, you need to carefully pick out the key information. After reading the information, ensure that you focus on the specific language in the final part of the question — this will contain the key words to ensure that you give a response which covers the information required in the mark scheme. For example, when discussing the reasons for a certain choice, you often need not only to make a decision about which choice should be made and why, but also explain why you rejected the other options. Do not just focus on the option you chose but also explain clearly why the other options were not suitable.

Core study skills

This section will focus on skills that are applicable to all written answers. The emphasis of the core skills section is on producing effective responses to short, structured questions, typically worth between 1 and 3 marks.

Typical command words used on these short, structured questions are:

→ State

→ Explain

→ Describe

Tell the 'complete story' in a logical sequence and don't forget to state the obvious.

Demonstrating factual knowledge

You will always obtain some marks in your examinations by accurately recalling specific information such as units and definitions. We have already discussed the reduction in the amount of marks available for recall of knowledge (AO1), but it still represents 33% of the A-level grade. Remembering information directly from the specification and then accurately reproducing this information without being asked to demonstrate understanding is classed as 'knowledge in isolation'. There is a limit of 15% of the marks available for 'knowledge in isolation'. At a maximum this could still yield around 40 marks across your entire A-level. These are the easiest marks to obtain and to reach the top grades you must succeed in getting these.

This shows the importance of learning standard definitions. The physical ideas you present should be complete and aligned to the number of allocated marks. This means that you need to use the number of allocated marks to determine the amount of detail you will include in your answer.

Worked example 3.1

Coherence

State what is meant by coherent wave sources.   (2)

Step 1: Read the question — identify the command word and marks allocated.

This is a '**state**' question, which is AO1: factual recall. i.e. this is a definition that should be learned prior to the exam. Many of these types of recall question appear on different past papers. You should ensure that you have a method of learning these key definitions, such as using flashcards or an app or piece of software to test yourself. Ensure that you include the full definition.

Coherence is related to the interference of waves. Interference patterns lasting long enough for us to see them can only occur if the superposing waves are coherent.

Step 2: Plan your answer.

There are 2 marks available, so the mark allocation will be as follows:

Sources are coherent if they have the **same frequency** ✓ [and since $I = f\lambda$ and speed is constant for a particular wave in the same conditions/material, then you could also say] or **same wavelength**. The waves must also have a **constant phase difference.** ✓

When the frequency is the same it also means that the wave is monochromatic. It is not enough for sources to just be monochromatic — for example, light from a red light may be of a single frequency, but it is incoherent because the photons of light emitted do not have a constant phase relationship. Thus the second marking point is essential in defining coherence. You should also note that even though a light may be 'red', it can consist of a range of wavelengths, so this would also make interference effects difficult to detect.

Recalling relevant information

In addition to learning specific definitions, there are also many descriptions, such as the effect of temperature on resistance, that you should learn. The added caveat with descriptions is that you must ensure that you are applying the description to the situation being asked about in the question, and not simply writing what was asked in a previous question you have answered. For example, if the question is asking you about the effect of temperature on resistance, it makes a huge difference whether the context is a metallic conductor or a thermistor. Not taking the specific situation into account will result in zero marks in many cases.

For these types of question it is important to first recall the relevant information regarding the physical principles involved, and then describe how the rule or phenomenon applies in the particular context of the question.

> ☑ **Exam tip**
>
> ### Use diagrams to help communicate your ideas
>
> The phrase 'a picture is worth a thousand words' can be helpful to bear in mind when questions require a longer descriptive response. Often a **well-drawn and clearly labelled diagram** can communicate your ideas more effectively than paragraphs of descriptive writing. Diagrams are particularly helpful when trying to understand the effect of forces on an object.
>
> When you are asked to add information to diagrams be sure to draw carefully and accurately.

Worked example 3.2

Describing using relevant information

A magnet is dropped vertically through a coil of wire, which is connected to a circuit. There is an ammeter connected to the circuit.

Describe what the student would see as the magnet passed through the coil and out of the other end. Analyse the differences, if any, if this procedure was repeated on the Moon. (5)

Step 1: Marks, Action, Physics, Situation (MAPS):

M: 5 — (distributed across both parts of the question — max 3 for either section).

A: '**describe**' — just say what you will observe; '**explain**', using your knowledge and understanding of the physics, how and why this changes on the Moon.

P: the underlying physics principles are: Faraday's law: ($\varepsilon = N\Delta\phi/\Delta t$) and Lenz's law (the direction of induced emf would cause a current to flow in a conductor, which would generate a field to oppose the change that created it).

S: magnet accelerating into and then continuing to accelerate out of a coil of wire. The procedure is then repeated on the Moon where acceleration due to gravity, g, is smaller. Therefore the magnet will experience a smaller acceleration.

Step 2: Describe how the current changes as the coil descends:

The current would increase, from zero, as the magnet approached the coil; as it reached to halfway through the coil, the current would reverse direction. ✓ The current in the reverse direction would be larger. ✓ As the magnet fell beyond the coil the current would reduce back to zero. The second increase in current would last for a shorter amount of time. ✓

(Max 3)

A diagram would arguably convey this description more effectively. Either of these two diagrams communicate the above sentence clearly, and leave less room for ambiguity.

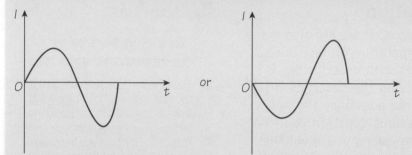

Figure 3.1 Two graphs showing current against time for a magnet falling through a coil of wire

Step 3: Explain the differences in the motion of the magnet on the Moon:

There is less gravity on the Moon. Therefore the magnet would not accelerate as quickly and so would fall at a slower speed through the coil. ✓

Step 4: Explain how the different motion changes the emf and current:

This means that the emf and therefore current induced would not be as high. The current would still increase and then decrease, but would not reach as high a value ✓ and the overall time from start to finish would be longer. ✓

(Max 3)

Using technical language

When communicating your ideas, technical language should be used whenever possible. Many candidates lose marks by not being precise enough with their technical language. Often students are deliberately vague because they do not have sufficient knowledge of the particular situation, but sometimes it is because they are not aware of how important it is to use the appropriate technical language at all times.

The difference between...

Explain the momentum change when an object rebounds elastically from a wall. (2)

Poor-quality answer	A/A*-grade answer
When the object rebounds from the wall its velocity changes, therefore since momentum = mass × velocity, the momentum also changes.	Momentum is calculated by: momentum = mass × velocity. Velocity is a vector quantity which means it has size and direction, which also means that momentum is a vector quantity. When the object rebounds, the direction reverses which reverses the sign of the velocity and also reverses, the sign of the momentum. Therefore change in momentum for elastic collision: $mu - -mu = 2mu$.
Zero marks: no mention of vector nature of velocity or momentum. Vague description of 'change'.	

Other areas where candidates typically lose out on marks

When you describe tensile stress, you need to write **force per cross-sectional area** rather than just saying force per unit area. When defining tensile strain, you have to specifically say extension per unit length rather than change in length per length.

When asked to explain the significance of the critical temperature of a material in relation to its resistance, many students will have a general understanding and will say that the material needs to be *at* the critical temperature in order to have zero resistivity and therefore zero resistance. This typical response is not precise enough: the material needs to be *at or below* the critical temperature in order to have zero resistivity.

Describing wave motion — it is important to understand that it is not the wave that oscillates parallel to the direction of energy transfer, but it is the **particles** that are oscillating.

Worked example 3.3

Faraday's law

State Faraday's law of electromagnetic induction. (1)

Step 1: **Read the question.**

This is a 'state' question so again it requires recall of a learned definition. You need to be precise with your statement to get the marks. For example, all of the following statements would score zero marks:

1 'The emf is proportional to the change of flux.'
2 'The emf is proportional to the rate of change of flux linkage.'
3 'The direction of the induced emf is in the direction to oppose the change in flux producing it.'

You should be able to see why these statements would score zero marks. We will go through this shortly, but first note that **many of the definitions asked for in written responses are given in a mathematical form on the examination equation sheet.**

Stating Faraday's law correctly is made much easier by referring to this equation sheet:

$$\varepsilon = \frac{N\Delta\phi}{\Delta t}$$

Step 2: **Write an accurate response.**

So 'the induced emf is equal to the rate of change of flux linkage' is a correct statement of Faraday's law. But you could also write that 'the magnitude of the induced emf is proportional to the rate of change of flux'.

Now we can look carefully at the three previous statements to assess their failure to be precise with technical language:

1   This statement does not include the term **'rate of change'** (of flux).
2   The emf is **equal to** the rate of change of flux linkage, not proportional to it.
3   This is a correct statement of Lenz's law, rather than Faraday's law. Faraday's law tells us that the magnitude of the emf is equal to the rate of change of flux linkage. Lenz's law tells us the direction to the induced emf: it produces induced currents, which produce magnetic fields that act to oppose the changes causing the initial change in flux linkage.

Use unambiguous language

This is related to the previous section on technical language, but deserves its own section. When you explain a phenomenon you should express your ideas clearly and unambiguously. This means that the examiner must be sure of what you mean when you are describing or explaining your ideas.

 Exam tip

When asked to suggest how a certain quantity will be affected by a change in another factor, you should never just state that it changes — this is an example of an 'equivocal' statement, which means that it is open to more than one interpretation. If possible, try to explain numerically how the quantity will change rather than give a vague description of decrease or increase.

The difference between…

(a) Explain the changes in energy and motion as an object falls.   (2)
(b) Suggest an improvement that could be made to this experiment.   (2)

Response 1	Response 2
(a) When the object's height decreases energy is converted, so the ball increases in velocity.	(a) As the object falls its gravitational potential energy store decreases and the kinetic energy store increases. Since $E_k = \frac{1}{2}mv^2$, the velocity also increases.
Response mentions energy conversion without specifying the types of energy involved.	*Response clearly states the changes in energy and relates these to the motion of the object.*
(b) Making a video would improve the accuracy.	(b) Record the motion of the pendulum using a video camera. By analysing the motion frame by frame you could reduce the uncertainty in determining the displacement of the pendulum compared with when it is swinging in real time.
This is just a general statement, like the similar go-to phrase 'to make it a fair test' that crops up when students don't want to analyse the situation in detail or consider a precise response.	*This not only suggests an improvement but, guided by the 2 mark allocation, explains why this procedure would improve the experiment.*

Higher-order study skills

Application

Candidates often lose marks when a question appears to be asking for a standard explanation for a process but, in fact, the question is asking you to **apply** physics to the context of the question and not just repeat a law or explanation from memory. When applying your knowledge and understanding in context, you need to carefully consider what conditions may have changed from the 'textbook' situation. You will need to explain these changes as well as their effect.

Worked example 3.4

Quantum phenomena: excitation

Explain the excitation of mercury atoms in a fluorescent tube. (2)

Although the question is short, many candidates will see the first three words and start writing: 'Explain the excitation'... However, bearing in mind the points from Chapter 2, you should read the question fully and navigate the question using MAPS to identify the key information:

Step 1: **Marks (M): 2** — needs a clear, concise response.

Step 2: **Action (A):** 'explain' — use your knowledge and understanding to give the reasons why the processes occur.

Step 3: **Physics (P):** excitation in atoms occurs when electrons absorb energy and move to higher energy levels.

Step 4: **Situation (S):** the key here is to focus your response on the context, i.e. do not just start explaining excitation, but do explain how mercury atoms are excited in a **fluorescent tube**.

Step 5: Write a clear, concise response:

Free electrons collide with electrons in mercury atoms ✓ moving them from the ground state to a higher energy level. ✓

Key features of an excellent extended response

Extended response questions are allocated 6 marks and give you the opportunity to show that you can produce a coherent and logically structured answer that links content from the specification through clear and fully sustained lines of reasoning. These types of question are also referred to as 'level of response' questions as you will be allocated to one of three levels depending on both the quality of your application of physics and your ability to communicate your reasoning.

The knowledge and understanding necessary to write an A*-grade answer to an extended writing question is called 'indicative content'. Mark schemes will provide a number of curriculum ideas covering the indicative content, but an A* grade needs more than just lists of indicative content. You can only obtain top marks by communicating your ideas effectively.

In order to reach top marks your answer must apply the following:

1 Be legible.

If the examiner cannot read what you have written you will score zero marks. Remember that the examiner will be trying to mark hundreds of papers and will be trying to do so as quickly (and accurately) as possible, but this means that they will not spend a long time trying to work out exactly what you have written: 'is that a 5 or a 3 or an 8?' They will not be able to give the mark if they cannot interpret your answer quickly.

2 Use appropriate grammar, spelling and punctuation.

The old SPAG (spelling, punctuation and grammar) marks have gone. (In previous versions of the specifications you could score 1 or 2 marks for simply writing with excellent spelling, punctuation and grammar — even if what you were writing contained incorrect physics!) Now, what matters is that the spelling, punctuation and grammar enable the meaning of what you are writing to be clear. To access the top level of answer you should only make one or two spelling or grammatical errors.

3 Use a style of writing appropriate to the question.

Different situations will require different styles of writing. This means that it may be more appropriate to use bullet points, e.g. in an equipment list. It may be appropriate to use an equation to show a particular relationship succinctly (remember to always define the quantities in any equation you state). It may be appropriate to use diagrams to communicate your point effectively, e.g. explaining the strong nuclear force (remember to include labels on all diagrams).

> **✓ Exam tip**
>
> The advice from examiners is that you should plan your response before you start to write. This will help you write your answer logically and with the minimum amount of repetition.

4 Be clearly organised and show linkage of ideas.

Your lines of reasoning should have a logical structure and be easy to follow. A key feature of an excellent answer is that it shows 'linkage of ideas'. This means that not only does it address content points from the mark scheme, but it does so in a way that relates them together to explain the bigger process. In practice, this generally means that the information should be presented in a chronological way. i.e. start at the beginning of the process and explain what happens until you arrive at the end of the process, do not jump back and forth in time, or back and forth between different ideas.

5 Be relevant.

Only include information that relates directly to the context of the question. Focus your answer on the specific information requested in the question.

6 Be coherent.

This means that you must be consistent in your reasoning and conclusions throughout your response. This is also referred to as 'fully sustained reasoning'. If you gain marks for a correct point, but then later on in your answer you contradict yourself, you will not only lose the mark for the content, but also lose the more holistic criterion of being coherent.

7 Be substantiated.

This means that you should provide sufficient evidence to support your ideas. Do not just say that some factor will increase. Explain why this will be the case using appropriate information to justify your assertion.

Worked example 3.5

Providing a coherent answer

Explain why the amount of gamma radiation decreases as the detector is moved further from the source. (1)

The intensity of gamma radiation obeys an inverse square law which means that it decreases exponentially. ✗

e The first part of the answer is almost enough to obtain the mark. It would be even better with a mathematical explanation e.g. $I \propto \dfrac{1}{d^2}$ where d is the distance between the gamma source and the detector. However, the candidate has described the reduction as exponential, this is not an inverse square law reduction — the answer is therefore not coherent and is awarded **zero marks**.

❗ Common pitfall

Don't waste time 'setting the scene' — focus on the relevant information.

Read the question carefully, as described in Chapter 2. Often in extended writing questions there are bullet points which identify the specific points that you need to address in your response. When planning what you are going to write ask yourself the question: 'Does this directly relate to what I am being asked to do (command word) and does it relate to the context of the question?'

Many candidates begin their answer by repeating the question — this gains no credit and wastes time.

Worked example 3.6

Electricity extended question

Students lack confidence in their understanding of the electricity topics and this is often evident in questions which require an extended written response involving the electricity area of the specification.

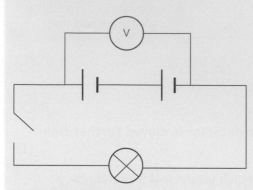

Figure 3.2 Circuit diagram

Figure 3.2 shows an electrical circuit with a filament lamp, battery, switch and connecting wires. A voltmeter of very high resistance is used to measure the potential difference across the battery terminals. All of the components are initially at room temperature.

Describe and explain what happens to the reading on the voltmeter when the switch is closed.
● **Consider the change as the switch is closed.**
● **Consider any further changes that may occur in a very short time after the switch has been closed.** **(6)**

Step 1: **Read** the question and identify relevant information.

'All components are initially at room temperature' — this is an unusual piece of information to include with a circuit diagram, but it implies that the temperature may not stay constant.

You need to **describe** what happens — the only effect we can see is the effect on the voltmeter, which measures the potential difference across the battery.

You also need to **explain** why any changes happen using your understanding of electrical circuits.

You need to address the bullet points to ensure you meet the marking criteria, so you must talk about the change as the switch is closed and then consider further changes that will occur in a very short time after the switch has been closed.

The question is worth 6 marks, so you need sufficient detail in your answer.

Step 2: **Plan** your answer.

A good answer will follow a logical sequence. In this case we will be looking at the pd value (this will be the description) at three different points in time: before the switch is pressed, just after the switch is closed and at a slightly later time. The answer will need to include an explanation at each stage to correctly identify why the changes are happening.

You need to think about what subject-specific content you can apply in this situation: *V*, *I*, *r* and *R*. Think about the link between current, potential difference and resistance, but also between current and temperature, temperature and resistance, and internal resistance and current.

Step 3: **Write** a clearly organised and coherent answer.

Before the switch is closed, no current is supplied, so the voltmeter will read the full emf, ε, of the battery. ✓ *As soon as the switch is closed, a current flows through the circuit and the battery loses volts because of its internal resistance. The reading of the voltmeter decreases because the pd across its terminals decreases from ε to $V = \varepsilon - Ir$* ✓ *where r is the internal resistance of the battery. As the current flows through the circuit, the filament bulb gets hot.* ✓ *Therefore the resistance of the circuit increases.* ✓ *This decreases the current in the circuit, which means the pd across the internal resistance decreases and the reading on the voltmeter will increase slightly* ✓ *(it will remain below the initial value, ε). The reading would then remain constant.* ✓

This is an excellent response. The candidate breaks the steps into three main stages: before the switch is closed, immediately after the switch is closed and a short time later. They apply their physics knowledge of internal resistance and potential dividers in a series circuit and combine this well with contextual information about the temperature. Technical language is used frequently and accurately throughout the explanation.

Synoptic assessment

There are three main types of question in your physics examinations:

1 The first type requires you to use your knowledge and understanding in one area of the specification, including relevant mathematical and practical skills.

2 The second type requires you to apply your mathematical skills and practical skills to unfamiliar areas of content.

3 The final type of question requires to make and use connections between different areas of physics. This is known as **synoptic assessment**.

The increased emphasis on higher-order thinking skills means that synoptic questions feature more heavily in A-level Physics. A deep understanding of underlying physical principles will enable you to analyse information and evaluate relationships between ideas from different areas of the specification. The linear course structure used by many exam boards also lends itself to synoptic assessment, as you should have relevant mathematical skills and subject knowledge to make links between the topics studied over the 2-year course. This is one of the key differences you will notice when you compare questions on the previous specifications with the current specifications. There was very little synoptic assessment in the previous specifications and this something you should be aware of when completing past papers as part of your revision.

The aim of synoptic assessment is to encourage you to see physics as a whole subject rather than as a series of unrelated topics. Synoptic questions require you to bring together your knowledge and understanding from more than one area of physics and apply them to a particular situation. Synoptic questions may also ask you to use your knowledge and understanding of physics principles to plan investigations or to analyse and evaluate experimental data.

Worked example 3.7

Synoptic question

Four particles approach a uniform electric field at the same velocity: parallel to the field lines, but in the opposite direction. Two of the particles are hadrons: one consisting of up–antiup, another of up, up, down. The other two particles are leptons: a muon and a positron.

This situation for one of the particles is shown in Figure 3.3:

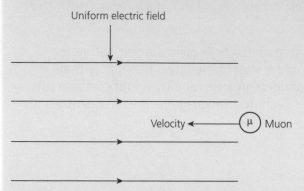

Figure 3.3 Particle in an electric field

Explain how you could identify the different particles by referring to the velocity with which they leave the electric field. (4)

This question brings together the topic of particle physics with fields, forces and motion. The basic principle is that electric field lines show the direction of the force on a positive test charge. (They do not necessarily show the direction of motion, just the direction of the force.)

With this in mind, we just have to understand that the particle will accelerate in the direction of this force. So really we just have to apply some information from the physics data and equation sheet to determine the charges of the particles involved — this process is even quicker if you can already recall this information.

> The up–antiup particle is a neutral pion, so has zero charge. This means that it will not experience any force from the electric field and will continue to move to the left with the same velocity as it entered the field. ✓
>
> The uud particle is a proton, so has a positive charge. It will experience a force to the right causing it to decelerate in the electric field. Thus it will leave the field with a velocity lower than its initial velocity. ✓
>
> The muon has a negative charge. Therefore it will experience a force to the left which will accelerate the particle. The muon will leave with a final velocity higher than its initial velocity. ✓
>
> The positron is a anti-electron, so it has a positive charge. It will experience a force to the right causing it to decelerate. Because it has a smaller mass than the proton, it will experience a greater deceleration whilst in the field and therefore will leave the field with the lowest velocity of the four particles. ✓

Worked example 3.8

Compare and contrast different quantum phenomena

This is a synoptic question because it draws information from different areas of the specification. It asks you to not only explain different phenomena, but it also requires the higher-order skill of comparing ideas to identify similarities and differences.

Compare and contrast the photoelectric effect, excitation and ionisation. (6)

Step 1: **Read the question and identify the mark allocation, action/command word, underlying physics principles and the situation/context:**

'Compare and contrast' means that you have to explain these phenomena in relation to each other, rather than in isolation.

This question does not have a particular context. The skill here is in showing an understanding of the similarities and differences between these different phenomena.

There are 6 marks allocated so it will be a level-of-response marking grid. Higher-level responses will have clear descriptions of the phenomena, but will focus on the similarities and differences between all three. This is what you are aiming for, so we do not need to consider the lower-level marking criteria.

Step 2: **Plan your response:**

In order to ensure a coherent and logical presentation of your ideas, it is a good idea to spend a few minutes thinking through the main components of your answer.

One idea that may help when comparing and contrasting ideas is to make a table:

	Photoelectric effect	Excitation	Ionisation
Electrons	✓	✓	✓
Photons	✓	✓	✓
Metals	✓	✗	✗
Atoms	✗	✓	✓
Emitting electrons	✓	✗	✓
Emitting photons	✗	✓	✗
Evidence for photon model of light	✓	✗	✗

A table like this clearly shows how the three different phenomena are related.

Step 3: **Write your response.**

The photoelectric effect, excitation and ionisation are all involved in the interaction between photons (particles of electromagnetic radiation) and electrons. The differences between them relate to the material involved: individual atoms or a metal surface as a whole, the particles providing the input energy and the type of particles emitted.

The photoelectric effect is the process where electrons are emitted from the surface of a metal when electromagnetic radiation above a threshold frequency is incident on the metal. The emission of electrons from the metal provides evidence for the existence of photons (particles of light) and it is the photon model which could then explain the existence of energy levels within atoms, which in turn explains excitation and ionisation. ✓

Excitation and ionisation involve electrons within individual atoms ✓ absorbing energy and moving to a higher energy level, whereas the photoelectric effect is about the electrons in the metal as a whole ✓. In ionisation the electron gains enough energy to leave the atom, but in excitation the electron remains in a bound state, before returning to a lower energy level. ✓ This is a key difference between excitation and the other two processes: a photon is emitted after excitation, whereas electrons are emitted in the photoelectric effect and ionisation. ✓ In all three processes the energy can be transferred to the electron by a photon: in the photoelectric effect this is the only way to input the energy, but in excitation and ionisation this energy can also be provided by collisions with other electrons. ✓

This response has attempted to put a structure to the arguments. The opening paragraph gives a general sense of the similarities and differences. The second paragraph then explains the photoelectric effect in detail and how it provides evidence for the mechanisms of the other two phenomena. The final paragraph is full of comparisons, with each statement providing more detail about how the three phenomena are related. An answer to this question is particularly difficult to structure, and without an initial plan it would be hard to produce a logical and coherent response.

Making judgements

Making judgements is one of the hardest skills in physics. This is because it uses all of the other skills as the basis on which to make your judgement: you have to read the information presented to you, then process, analyse and evaluate it in the context of the question, using relevant knowledge and understanding from the specification, and then, finally, you can make a judgement.

It is difficult to make judgements easier. One way to do this is to practise making judgements, by constantly evaluating and comparing different ideas as you come across them in your course. The other way to develop this skill is by practising examination questions.

Since the judgement you make will be based on the analysis and evaluation you have carried out, your final judgement will be assessed on whether it is logical and in line with the reasoning you have presented in the rest of your answer. If you have made some mistakes in your analysis, but your judgement is coherent with this analysis (even though it may be flawed) you will still be awarded the marks. When answering a question that asks for a judgement to be made, you must make a clear final statement where you reflect on and compare any values you have calculated.

As part of making your judgement you may be asked to evaluate risks and benefits of certain technologies, based on your scientific understanding. You may be asked to consider ethical issues in the treatment of humans, animals and the environment, or in business. You can also be asked to evaluate the ways in which society can utilise science and scientific thinking to impact on decision making in business or government.

Worked example 3.9

Judgement

The following question asks you to evaluate and make a judgement about a new piece of technology. Candidates find this type of question challenging. You are presented with lots of new information. You need to process this and use it to evaluate some of the new information against criteria stated in the first part of the text. Finally, you can come to a final judgement.

The inventor of a device called a mass–energy generator, which operates using Einstein's mass–energy equivalence principle, hopes to be able to patent their invention. In order to be granted a patent the invention must be all of the following:
- **something that can be made or used**
- **new**
- **inventive — not just a simple modification to something that already exists**

The device works as follows: an object of mass, *m*, is allowed to fall a distance, *d*, within the machine. As it falls the kinetic energy store of the mass increases. When it reaches the bottom of the device, the mass is converted into pure energy in the form of photons. The photons are reflected back up a distance, *d*, to the top of the device. When they reach the top section the photons are converted back into a mass and the process is then repeated.

Evaluate this patent application and decide whether it could be granted. (4)

Step 1: To grant a patent you must use the criteria given in the question.

The first criterion is that the device can actually be made, i.e. will it work?

Step 2: You must then evaluate the criterion by examining the operation of the device and applying your understanding of the physics involved.

As the mass falls, it gains kinetic energy. If the total energy is then converted into photons and reflected upwards, there will be the energy from the original mass plus the extra kinetic energy gained as the mass fell. This would be converted into a mass larger than the original mass, which means there would be even more kinetic energy to be converted into photons the next time around. The energy would be increasing indefinitely at each cycle. This does not comply with the law of conservation of energy and so this process cannot actually take place.

Step 3: You then need to make a judgement to decide whether the patent can be granted.

The second two criteria may, or may not, be true, but the first criterion is not satisfied, so the patent cannot be granted.

✓ Exam tip

Longer question techniques when you don't have background knowledge on the topic

There may be occasions when you are asked to describe and explain processes when you don't know any specific examples about the topic. In such cases you need to use the information provided in the question to help you structure your answer and give the required detail to maximise your marks. Think about what basic knowledge you have about the underlying physics principles that you can apply to this situation. Making a quick plan before you begin your answer can really help. The plan will enable you to identify gaps or inconsistencies in your reasoning, which you can then amend before writing your final answer.

Application to the exam

This question is about wave–particle duality and the nature of scientific research.

The Nobel Prize in Physics is awarded each year to celebrate the most outstanding contributions to the field of physics. By the time of the first award of the Nobel prize, evidence from experiments by Thomas Young and Christian Huygens suggested that light was a wave.

Further research into the wave and particle nature of light and matter has resulted in Nobel prizes being awarded to various other physicists:

Date	Scientist	Reason for Nobel prize award
1906	Joseph Thomson	Particle nature of electron
1921	Albert Einstein	Particle nature of light
1937	George Thomson	Wave nature of electron

Use the information presented in this question to evaluate the way in which society uses science to inform decision making.

Your answer should address:

→ the evidence supporting the different models

→ the role of the scientific community in validating new knowledge

→ how society uses science in decision making

You do not need to include the details of particular experiments. (6)

B-grade answer

The question says that the first model for light was that it was a wave. This is from experiments carried out by Huygens and Young. Young's two-slits experiment is when laser light is shone through two slits and the light produces a series of light and dark fringes on a screen a distance, D, from the slits. The width of the fringes, w, is given by $w = \frac{\lambda D}{s}$, where s is the separation of the slits and λ is the wavelength of the light. Then JJ Thomson proved that electrons were also a particles. Then Albert Einstein showed that light was a particle because of the photoelectric effect. This is wave–particle duality: light can be a wave and a particle, but it depends on the experiment. The different scientists kept investigating and carrying out experiments which revealed different aspects of nature. Scientists must carry out these investigations and if the new results are not explainable by the old theories then they have to come up with a new model, e.g. light can be a wave or a particle. The government, businesses and other organisations can use the scientific ideas to draft laws and improve technology because the conclusions from the experiments are valid.

Marks awarded: 4

This candidate's response describes evidence for three of the four phenomena discussed in the question (there is no evidence presented for electrons as waves). A lot of information is given at the beginning of the response which although correct, does not address the question specifically and gives a lot of information, e.g. about two-slit interference with light. The candidate has explained the basic idea (from GCSE level) as to why theories need to change if results from new experiments are not explainable with current theories. There is a brief link to how society can use science in a range of different areas.

A/A*-grade answer

The information from Nobel prize awards shows that our understanding of light and matter has changed over time — the theories change as a result of reproducible evidence from experiments.

Initially interference effects led scientists to believe that light was a wave. JJ Thomson then used electric and magnetic fields to show that negatively charged particles were emitted from a cathode. Einstein explained the photoelectric effect by concluding that light must consist of particles called photons. George Thomson then found that electrons produce interference patterns and so must also behave like waves.

This means that there is evidence for the wave nature and particle nature of light (electromagnetic radiation) and electrons (matter). Scientists use 'peer review' to ensure that new experimental data are valid and reproducible. The experiments must be reproducible, which means that other scientists must be able to obtain the same results using different apparatus or a different method. Once the integrity of the new results is ensured, a new model can be developed to explain the new data.

This means that society in general can trust science and use it to develop policies and laws, e.g. vaccinations, emissions laws etc. However, there are still areas where science does not give clear guidance and the data can be used by either side of an argument (e.g. global warming).

Marks awarded: 6

This candidate's response is well written: it is coherent, logical and even has a good introductory paragraph outlining the general direction of the rest of the answer. This shows that a plan was drawn up before the candidate began writing. The evidence for the wave–particle models is accurately and succinctly described. And there is effective linkage of ideas between the evidence, why we believe the evidence, and how this relates to wider society.

You should know

> Responses to short questions assessing your ability to demonstrate knowledge and understanding should use clear, concise statements.

> Use precise technical language whenever possible.

> Present your arguments clearly using unambiguous, unequivocal language.

> The style of writing should be appropriate to the ideas and information you are trying to communicate. Labelled diagrams and bullet points are sometimes more effective than very long descriptions.

> Extended-response questions require clearly organised, substantiated and coherent arguments.

> Practise making judgements, by constantly evaluating and comparing different ideas as you come across them in your course. This will also develop your understanding of the interrelations of different topics and develop your ability to answer synoptic questions.

> When answering a question that asks for a judgement to be made, you must make a clear final statement in which you reflect on and compare any values you have calculated.

4 Practical skills

Learning objectives

> To become familiar with the assessment methods for practical skills

> To develop the skills necessary to meet the standard for the practical endorsement

> To develop confidence with the key terminology of practical physics

> To develop an understanding of many key practical techniques, especially relating to improving the accuracy of investigations

> To apply your knowledge of the fundamental principles of practical physics to plan effective investigations

> To develop skills of analysis in the context of practical physics

> To develop the skills to evaluate investigations in terms of quality of results and experimental procedures

> To be able to draw conclusions and to suggest appropriate improvements to experimental design

Introduction

Physics is a practical subject, which means that physicists find answers to questions by observing and experimenting. You need not only the technical knowledge and skill to carry out experiments, but also the higher-level skills of evaluating procedures, suggesting improvements and analysing results to arrive at sensible conclusions.

You should approach every practical activity with a critical eye, asking questions such as: How can I improve my methodology? How certain am I about my conclusions?

Assessment of practical skills

Practical skills assessment in A-level Physics consists of two components:

→ **Direct assessment** during your practical activities. (The WJEC exam board also includes a practical exam — for more details, see Chapter 6.)

→ **Indirect assessment** through written examinations.

Direct practical assessment: the practical endorsement

The assessment of practical skills is a compulsory requirement of all A-level Physics qualifications. All examination boards have the same arrangements for assessing practical skills through something

Take it further

Bad Science by Ben Goldacre is a book which reveals how dangerous it can be to take at face value 'scientific' claims by various companies. It highlights the true value of the scientific method and good science.

called the Common Practical Assessment Criteria (CPAC). The CPAC mean that if you demonstrate the required standard to your teacher over the 2-year A-level course, you will receive a 'pass' grade on your certificate alongside your overall grade from the written papers.

To obtain the 'pass' grade for the practical endorsement, you must show mastery of five main practical skills to your teacher. To show mastery means that your teacher can confidently state that you have provided evidence that you can consistently and routinely show the following practical competencies:

→ **Follow written procedures** — you demonstrate that you can correctly follow instructions to carry out practical techniques.

→ **Apply investigative approaches and methods** — you select appropriate equipment and measurement strategies, take into account control variables, carry out techniques methodically and make adjustments to procedures as necessary when practical issues arise.

→ **Use a range of practical equipment safely** — identify hazards, assess risks and make adjustments as necessary.

→ **Make and record accurate observations** — obtain sufficient, relevant, accurate and precise data, and make records with appropriate units and conventions.

→ **Research, reference and report** — use appropriate tools and/or IT to process data, carry out research and report findings and make appropriate citations.

You can demonstrate these competencies in any practical activity which you carry out through the course, but each exam board has a set of 'required practical activities' which you must carry out. These 'core practicals' have been designed to give you practical opportunities to demonstrate the practical skills.

Indirect practical assessment: questions in written exams

Your practical skills are also tested in the written examinations. This is called indirect assessment. At least 15% of the overall assessment of A-level Physics will assess knowledge, skills and understanding in relation to practical work, which will be based around the required practicals.

There are four main areas which will be assessed in the examinations:

→ **Planning** — design experiments to solve problems in practical contexts, identify variables that must be controlled and evaluate experimental methods against whether they will meet expected outcomes.

→ **Implementing** — explain how a wide range of practical apparatus and techniques should be used, decide on appropriate units for measurements and present observations and data in appropriate ways.

→ **Analysis** — process and interpret experimental results using appropriate mathematical skills to analyse quantitative data. Use significant figures appropriately. Plot and interpret graphs including measurement of gradients and intercepts.

➜ **Evaluating** — draw conclusions, identify anomalies, determine limitations of experiments and suggest improvements in terms of procedures and apparatus, consider margins of error, accuracy and precision of measurements, including an assessment of percentage uncertainties.

The four stages above are the natural way in which you would go about carrying out an investigation, starting with planning, then implementing that plan, analysing the data and then evaluating the results to draw conclusions and present a confidence level about these conclusions.

Core study skills

Key practical terms

Each examination board produces a set of key practical terms developed from *The Language of Measurement* by the Association for Science Education. It is essential that you have complete recall of these terms as they are often used imprecisely and interchangeably by candidates. Chapter 5 gives advice on how best to learn and retain this kind of information.

> ### Activity 4.1
>
> Look up the key practical terms for your examination board and ensure you learn the definitions. Practise using these terms whenever you are carrying out practical activities or data analysis by referring to as many of them as possible in your lab book.

> ### ✓ Exam tip
>
> If the independent variable is continuous, you should produce a line graph of your results. If the independent variable is categoric, you should use a bar chart to analyse your results.

> ### Worked example 4.1
>
> #### Introducing key terminology
>
> This example shows the key practical terminology being used in the context of planning an investigation.
>
> **You have been asked to plan an investigation into the factors that affect the frequency of stationary waves on a string.**
>
> The task asks you to measure the effect on the 'frequency of stationary waves', so 'frequency' is your **dependent variable**. There are many other variables in the experiment: length of string, tension on string, mass per unit length of string — which may depend upon the diameter of the string and the material from which it is made. The type of material is a **categoric variable**, but all of the others are **continuous** (as they are numbers).
>
> You must carry out a systematic investigation, so you should only change one variable at a time (the **independent variable**) and keep the others constant (the **control variables**). You can select any of the other variables to investigate first, e.g. the length of the string. The control variables, which must remain constant, would then be the tension on the string and the mass per unit length of the string.
>
> You could then extend the investigation by changing the independent variable and exploring how a different factor affects the frequency. For example, you could make the tension the independent variable and see how it affects the dependent variable (frequency). You would need to keep the other variables controlled, e.g. length of string and mass per unit length. By

allowing only one variable to affect the dependent variable each time, you will generate **valid** data from which to draw conclusions.

When describing procedures such as this in an exam, ensure you specify the exact **range** of the independent variable and how you will ensure the other variables remain constant. Also state what **apparatus** you will use to measure the independent and dependent variables along with the **resolution** (the smallest measuring interval on the instrument).

Planning

Your first task is to decide what problem you are going to solve, and determine the aim of your investigation.

Measurements

Use the experimental aim to determine the independent, dependent and control variables and then decide on the **range** and **interval** of the measurements.

Obtaining an adequate range of the variable should be your first consideration. Then you can decide on the interval.

Range

You should maximise the range of your independent variable so that you can see how far the pattern or relationship between the variables extends. The limits you choose for your range should be due to difficulties in obtaining data beyond these values, rather than due to time constraints.

Interval

Random errors are present each time a measurement is made, but the effect of these random errors can be reduced by making more measurements and calculating a mean.

There are three ways to take more measurements:

→ **Repeat readings**: take multiple measurements of the dependent variable for the same value of independent variable.

→ **Increase the range of the independent variable**: this will lead to increased data collection, especially if the interval remains constant.

→ **Use a smaller interval**: this will generate more data, so that the line of best fit can be drawn more accurately. The smaller the interval the more sensitively you will discover the changes in the dependent variable. *You do not need to use a constant value for the interval through the entire range of readings.* Where there is a sudden change in your dependent variable, it would be instructive to use a smaller interval in this region to examine what is going on in more detail.

> ✅ **Exam tip**
>
> Read the question carefully. Pay special attention to the definition of the letters given in the question, as these often represent different measurements and will be different for each practical set-up.

Worked example 4.2

Measurements

Discuss the issues when deciding upon the range of your variables when investigating the oscillations of a mass–spring system. **(2)**

Your independent variable is the mass on the spring and the dependent variable is the time period of the oscillations. You should use as big a range of masses as possible: the upper limit for the mass is decided by the point at which the spring starts to deform ✓ (perhaps around 1 kg). When the mass is too small the oscillations can be erratic, which leads to difficulties in determining a time period when timing for multiple oscillations. ✓

Instruments and experimental design

The key criterion when considering which measuring instrument to use concerns the resolution of the device. The rule is that the **minimum resolution must enable changes in the dependent variable for each change in the independent variable.** The measuring device is not the only thing contributing to inaccuracies. You also need to consider techniques and procedures. A greater resolution will tend to produce a smaller percentage uncertainty in the reading for that quantity (see the section on uncertainty on page 72).

When carrying out practical work, you should continually ask why the particular experimental methods and apparatus are being used, and what the effect might be of changing a particular aspect of the method.

Worked example 4.3

Planning

You are investigating how the interference fringe spacing from a double slit changes with distance from the screen. Discuss the methods you could use to ensure the double slits and the screen are aligned correctly. **(4)**

Use MAPS to navigate the question:

M: — 4

A: — 'Discuss' you need to explore a range of different ideas and their effects on the factors involved.

P: — From the equation, the fringe spacing, $w \propto D$, the distance between the slits and the screen.

S: — The distance needs to be able to increase whilst keeping the slits parallel to the screen. Ideally the apparatus needs to be able to be moved easily whilst maintaining the correct alignment.

A simple method uses a metre ruler to measure the distance between the two slits and the screen at each side. To ensure the slits and the screen are parallel the two distances must be equal. ✓

A different method uses two set squares and a metre ruler. First, place a set square against the slits. ✓ Place a metre ruler perpendicular to this set square. You can then place a second set square at any distance along the ruler, ensuring that the edge is perpendicular to the metre ruler. If you align the screen against the second set square then it will be parallel to the slits. ✓

This method has the advantage over simply measuring the distance at each end because it allows the distance to be varied continuously whilst they remain parallel to each other. ✓

e Two set squares can be used to check relative alignment, either ensuring objects are perpendicular or parallel. This response describes two different methods and makes a judgement regarding which is the more effective one to use in this situation.

☑ **Exam tip**

The key points regarding selecting and using equipment are related to being specific about the instrument and exactly how it will be used. For example, saying you will 'use a ruler to measure the distance' is not enough to guarantee the mark. You would need to specify carefully how the ruler readings will be used and how the ruler and the object will be aligned.

Implementing

Key apparatus and techniques

Presenting data and observations in an appropriate format
The following guidelines should apply to your practical work:

1 **All observations and raw readings should be recorded.**
2 **Raw readings should be recorded to the resolution of the measuring device.**
3 **Results should be recorded in a results table.**

(a) Results for the **independent variable** should always be placed in the first column of the results table.

(b) The **title** of each column should be clearly labelled with two things:

(i) The **physical quantity**. This is often just a symbol, e.g. 'T' for time period or 'd' for distance. Take care to use the correct case, e.g. l instead of L.

(ii) The **unit** preceded by a solidus (forward slash) e.g. T/s, or d/mm. Although using a solidus is the preferred format, you could also write 'Time in s' or 'Time (s)'.

Worked example 4.4

Length measurement

Using a standard metre ruler, you measure a length of string as 80 cm. How should you write down your measurement?   (1)

The length should be recorded as: 800 mm ✓.

You could also write 80.0 cm or 0.800 m, but not 80 cm. A standard metre ruler has 1 mm graduations, which means it has a resolution of 1 mm.

Worked example 4.5

Determining instrument resolution from data

The mass of a typical 100 g slotted mass is recorded as 0.10 kg. Comment on this result. **(2)**

> The value of 0.10 kg says that the mass was only measured to a resolution of 10 g. ✓ However, the typical top-pan balance has a resolution of 0.01 g, so the recorded figure should show this by being written as 0.10000 kg. ✓

The number of decimal places you record your data to is important because it gives information about the resolution and therefore can relate to the accuracy of your final conclusions.

Worked example 4.6

Techniques to improve accuracy

Explain the techniques you could use to increase the timings when investigating oscillations. **(3)**

> Use of fiducial marker at the centre of the oscillations as this precisely identifies the beginning and end of the cycle. This is because the object is travelling fastest at the centre, making it easier to identify the exact time at which it passes the marker. ✓
>
> Time over multiple oscillations. Try to time at least 20 oscillations of the object. But be aware that you don't need to use a fixed number of oscillations, when the time period is shorter it may be necessary to time over more cycles, and if there is heavy damping the amplitude may have reduced to zero before 20 oscillations. Maximising the total length of time you measure reduces the percentage uncertainty in the time period. A longer time measurement also reduces the effect of human reaction time. ✓ Take care not to miscount the number of oscillations — say 'nought' as you begin the timing.
>
> Allow transient oscillations to die down before starting to time the motion. Initial oscillations of any vibrating system being acted on by a driving force are irregular and have a varying amplitude and time period. ✓ The transient oscillations gradually move to a steady state where the driven oscillator has the same frequency as the driving force (which can be different from its own natural frequency). The speed at which the transient oscillations give way to steady oscillations depends on the degree of damping in the system, but typically you can start timing after one or two oscillations.

Saying 'use a data logger' is not sufficient to obtain a mark. Data loggers are often the best devices to use in experiments that require data to be collected over a very short or a very long time scale, such as fast motions and capacitors charging or discharging. You also need need to describe what type of sensor you will use, what measurements it will take and what sample rate you will use.

> **✓ Exam tip**
>
> All experimental readings should be repeated and a mean calculated. This reduces the effect of random errors and allows anomalies to be detected which can then be discarded (before averaging).

Worked example 4.7

Measuring lengths

Lengths of objects can be measured with vernier calipers, micrometers and rulers. Discuss the best choice of instrument to measure a length of 8 cm. (3)

The micrometer does not have a large enough range to be used in this situation. ✓

Measuring an 8 cm length with a ruler will produce a percentage uncertainty of $\frac{1}{80} \times 100 = 1.25\%$ ✓

The same reading carried out using vernier callipers would produce a percentage uncertainty of $\frac{0.1}{80} \times 100 = 0.125\% = \pm 0.13\%$.
So the best instrument to use would be the vernier callipers because they have the smallest percentage uncertainty for the length measurement. ✓

The table below shows the trade-off between higher resolution and smaller range:

Instrument	Resolution	Maximum range
Metre ruler	±1 mm	1000 mm
Vernier callipers	±0.1 mm	100 mm
Micrometer	±0.01 mm	30 mm

Activity 4.2

Reading a vernier scale

A vernier scale is a second set of scale markings which help you determine more accurately a measurement that falls between two graduations on the main scale.

Use the example in Figure 4.1 to help read the scale:

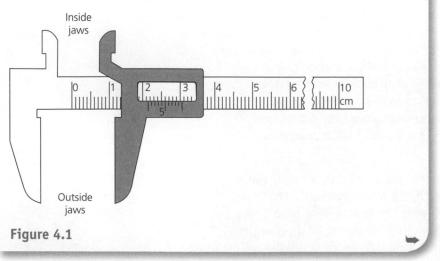

Figure 4.1

How to read the vernier scale:

- Use the main scale to find the nearest whole number and first decimal — 2.1 cm (21 mm) in the diagram.
- The point at where the lines on the main scale are exactly lined up with those on the vernier scale tells you the next decimal place — in the example, this is at '5' on the vernier scale i.e. 0.05 cm
- Add the second reading to the first to determine the actual measurement. The reading in the example is 2.1 + 0.05 = 2.15 cm (21.5 mm).

What is the reading on the vernier scale in Figure 4.2?

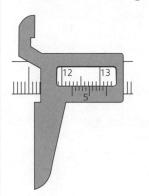

Figure 4.2

Answers online – see page 3

Answers online – see page 3

> ### ✓ Exam tip
>
> **Reading a micrometer**
>
> Before taking a measurement with a micrometer you should check for a zero error by closing the jaws with no object between the jaws.
>
> - The jaws should be closed using the ratchet until you hear a click.
> - It is important that you do not over-tighten because it may give an inaccurate reading by squashing the object and/or damaging the thread on the micrometer.

> ### Activity 4.3
>
> Look up '12 Physics apparatus and techniques' using either your specification or practical handbook. Make sure you have evidence of applying these skills in your lab book.

Higher-order study skills

Analysis

Once the experimental plan has been implemented, the next step is to process the collected data, which may involve further calculations, plot the data and interpret the results.

Percentage difference

A very simple way to analyse an experiment result is to use the **percentage difference**.

$$\text{percentage difference} = \frac{\text{your value} - \text{accepted value}}{\text{accepted value}} \times 100$$

Graphs

Plotting and interpreting graphs

Rules for plotting graphs

You will often have to process at least one of your measured quantities before plotting on the appropriate axes.

1 Use sensible scales.

Points should cover at least half of the grid horizontally and half of the grid vertically. If necessary, a false origin should be used to meet these criteria, and this must be clearly marked.

You should use a linear scale with a multiple of 2, 5 or 10. Choosing different scales will lead you to having to estimate or use a

> ### ✓ Exam tip
>
> Percentage difference is the easiest way to check 'proof of claim' type questions where you are asked to check the extent to which a certain claim is true:
>
> - If the experiment has produced a valid result with a percentage difference that is 5% or less of the value in the question, then you can accept the claim.
> - If the percentage difference is larger than 10%, the claim should be rejected.
> - The grey area is between 5% and 10%, where you will have to decide on a case-by-case basis whether to accept the claim or not.

calculator to determine the positions of points. Different scales are more likely to lead to plotting errors and often lead to errors when determining the gradient of the line of best fit.

2 Label the axes.

You should ensure both axes are labelled with units.

3 Plot the points accurately.

Each point should be plotted to an accuracy of ±1 mm. Double check any anomalous points.

4 Draw a line of best fit.

There should be an equal number of points above and below the line of best fit. All of your data should fall within 2 mm of the line of best fit.

5 Calculate the gradient and determine the intercept.

When calculating the gradient of a graph you should show all of your working clearly on the graph. Your y-step and x-step should both be at least eight semi-major grid squares. Aim for at least an 8 cm by 8 cm triangle on your graph.

There is lots more information on analysing graphs in Chapter 1.

> ✓ **Exam tip**
>
> When calculating the gradient of a graph do not forget to include any standard prefixes or other multiplication factors, e.g. KN, μm or × 10^{14} Hz, in your calculation.

Activity 4.4

Sketch graphs for the following mathematical relationships (where k is a constant):

(a) $y = \dfrac{k}{x}$

(c) $y = \dfrac{k}{x^2}$

(e) $y = \sin x$

(g) $y = e^{\pm x}$

(i) $y = \cos^2 x$

(b) $y = kx^2$

(d) $y = kx$

(f) $y = \cos x$

(h) $y = \sin^2 x\backslash$

Answers online – see page 3

Worked example 4.8

Analysis using graphs — direct proportion

A student says that her experiment shows that the independent and dependent variables are directly proportional. Explain using a results table and the graph how the student made this judgement. (3)

To make the judgement using only data from the results table she would need to determine whether when the value of the independent variable is doubled, the dependent variable is also doubled. ✓ There are two features of a graph for data that are directly proportional: the line of best fit is a straight line ✓, through the origin ✓.

Too many candidates remember the 'passes through the origin' part, but fail to explain that the graph must also be **linear**. For two quantities to be proportional the line must be linear.

Showing relationships between variables

You are often asked to justify whether experimental data can be explained by a particular equation, i.e. does the equation correctly describe the relationship between variables in the investigation.

Inverse proportion

To show that two quantities are inversely proportional it is not sufficient to say that as one quantity increases the other decreases.

If two quantities x and y are inversely proportional we would write:

$$y \propto \frac{1}{x}$$

Mathematically we can say then that $y = \frac{k}{x}$, where k is a constant. This means that $xy = k$. Essentially this means that x multiplied by y is equal to a constant. This is extremely useful in practical physics as a way of determining whether two quantities are inversely proportional.

Inversely proportional means that by multiplying both quantities together they will be equal to a constant, i.e. xy = constant.

Worked example 4.9

Significant figures

Calculate the speed of an object that takes 6.2 s to travel 448 mm. (2)

Step 1: Identify the appropriate equation:

$$speed = \frac{distance}{time}$$

$$s = \frac{d}{t}$$

Step 2: Substitute values into the equation:

$$s = \frac{d}{t}$$

$$s = \frac{448 \times 10^{-3}}{6.2} \checkmark$$

$$s = 7.226 \times 10^{-2} \, ms^{-1}$$

Step 3: Write your final answer to an appropriate number of significant figures. The distance is given to three significant figures and time is to two significant figures, so the result for speed should be quoted to two significant figures.

$$s = 7.2 \times 10^{-2} \, ms^{-1} \checkmark$$

Two significant figures is correct in this example, but what if, as will more frequently be the case, the raw time from the stopwatch is given as 6.20 s. If the stopwatch was stopped by an experimenter, then you need to take their reaction time into account, so the real time is actually 6.2 s (as stated in the previous question).

Also, what if the value for distance had been quoted at 450 mm? Is this to two or three significant figures? When tackling the practical examination questions you need to apply slightly different logic to the data. In a theory paper it would be expected that 450 mm was to two significant figures. However, *in a practical context, you need to consider the measuring device*. A length that large would be measured with a ruler, which has a resolution of 1 mm. So you know that the quantity must have been 450 mm rather than 449 mm or 451 mm. Therefore, it has been quoted to three significant figures. In fact the only way to remove this kind of equivocation is to use standard form: 4.5×10^{-3} m would be to two significant figures and 4.50×10^{-3} m would be to three significant figures.

Worked example 4.10

Testing relationships

Two equations have been proposed to explain a set of experimental data: $A = kB$ and $A = m\sqrt{B}$, where k and m are constants.

Outline the analysis you would perform on the data to show which equation correctly describes the data.   (3)

If you have data in a table you could calculate values of $\frac{A}{B}$ and $\frac{A}{\sqrt{B}}$ for multiple values. If one of the ratios gives a constant value each time, then that equation explains the experimental data. ✓

Calculate the % difference using $\frac{\text{value 1} - \text{value 2}}{\text{mean value}} \times 100\%$. If there is less than a 5% difference you can consider the values to be constant. ✓

You could also plot a graph of A against $\frac{1}{B}$ and a graph of A against $\frac{1}{\sqrt{B}}$. If one of these gives a linear graph, then that equation explains the experimental data. ✓

Evaluating

The final practical skill of evaluation requires you to assess the quality of your results and draw sensible conclusions. You will need to evaluate the validity, precision, accuracy and uncertainty of your measurements. You will be expected to identify anomalies and limitations in the experimental procedures as well as suggesting improvements to the experimental design. In reality, the evaluation forms a feedback loop with the planning stage, where the refinements to procedures will be incorporated into subsequent experiments to obtain more accurate data.

Uncertainty

Every measurement has an uncertainty. Even the 'constants' from the data sheet, such as the charge on an electron or the speed of light in a vacuum, have an uncertainty and are not the actual true value. Every experimental measurement we have ever taken is simply the closest that the particular experimental set-up could obtain using the particular apparatus and techniques. This means that a measurement result is only complete if it consists of a value and a statement about the uncertainty in the value.

A good experiment will use appropriate techniques and equipment to minimise the uncertainty, and provide a realistic calculation of the uncertainty, such that the true value of the quantity being measured lies within the range specified by the uncertainty. The degree of uncertainty will therefore depend upon:

→ the instrument(s) being used to make the measurement
→ the method, i.e. the way in which the measurement is made
→ the quantity measured, which may not be constant

Uncertainties in readings and measurements

The uncertainty in a **reading** is no smaller than plus or minus half of the smallest scale division. A reading is found from a single judgement using a piece of equipment, e.g. determining the temperature using a thermometer, or the mass of an object using a top-pan balance.

A **measurement** is defined as being the difference between two readings. A measurement thus includes the uncertainty associated with two readings, which means it has twice the uncertainty of a single reading. A measurement is where the value is the difference between two judgements, e.g. measuring length using a ruler, or an angle using a protractor.

This gives the absolute uncertainty for a quantity when you have only one value, or when all of the repeats give the same value.

Worked example 4.11

What is the uncertainty in your temperature reading using a thermometer with 1°C scale graduation? How does this change when using 0.5°C graduations? (2)

When using a thermometer with 1°C scale graduations the uncertainty will be at least ±0.5°C. ✓ If the graduations were 0.5°C, then the uncertainty in the temperature reading would be ±0.25°C. Since you cannot read to a hundredth of a degree, this should be rounded to ±0.3°C. ✓ For a top-pan balance with a resolution of 0.01 g the uncertainty is ±0.005 g for each reading.

Annotated example 4.1

Explain how the uncertainty of a length measurement can be reduced when using a ruler. (3)

This 'initial value uncertainty' occurs whenever the experimenter can set the zero incorrectly, but does not apply to equipment such as top-pan balances or thermometers.

A standard ruler has a resolution of 1 mm. There is an uncertainty of ±0.5 mm at the zero end of the ruler and an uncertainty of ± 0.5 mm at the other end of the ruler where the reading is taken. ✓ The total uncertainty in the length measurement is 0.5 mm + 0.5 mm = ±1 mm. ✓ If the experimenter is able to fix one end of a measuring scale ✓ then the lengths that would have been measurements (taking into account an alignment uncertainty at both ends) become readings, with only an alignment uncertainty at the point where the reading is taken. This technique halves the uncertainty to ±0.5 mm. ✓

This is the minimum uncertainty associated with this measurement and so should be stated as: 'at least ±1 mm'.

Often, the way an instrument is used will determine whether the experimenter is taking a reading or a measurement, and this will have an effect on the final uncertainty.

Worked example 4.12

A voltmeter capable of reading up to 100 V, displays a reading of 14.2 V. Criticise the statement: 'The voltmeter has an uncertainty of 0.1 V.'   (3)

When reading a digital scale you usually say that the uncertainty is ±1 in the least significant digit. In this case the least significant figure is the 0.2 V so the uncertainty of this reading is ±0.1 V. ✓ The voltmeter has a large range, and will therefore probably have different range settings. On a smaller range setting, e.g. 0–20 V, the uncertainty could be as already stated: ± 0.1 V, but on a higher range setting, e.g. 0–100 V, the uncertainty of the voltmeter may be ±1 V. ✓ The statement correctly states the magnitude of the uncertainty on one particular range, but omits the ±, which means that the true value could be greater or less than the value displayed on the screen and also considers the uncertainty to be fixed throughout the range when this may not be the case. ✓

Determining absolute uncertainty

When there are multiple readings for a quantity the absolute uncertainty is estimated using:

(absolute) uncertainty = ±0.5 × range

You would then quote the value of the quantity as: mean value ± uncertainty.

Percentage uncertainties

It is useful to be able to convert uncertainties into percentage uncertainties, as they allow you to compare which aspect of the procedure contributes more to the overall uncertainty.

$$\text{percentage uncertainty} = \frac{\text{uncertainty}}{\text{value}} \times 100\%$$

Combining percentage uncertainties

When data are processed this often involves calculating quantities by combining different measurements together. The following rules allow you to calculate the combined uncertainty:

→ If quantities are added or subtracted, add the absolute uncertainties of each quantity.

→ If quantities are multiplied or divided, add the percentage uncertainties of each quantity.

→ If a quantity is raised to a power, then multiply the percentage uncertainty by the power.

Using uncertainties to evaluate experiments

If the true value or hypothesised value is within the range specified by the uncertainty on the final value determined by the experiment, then the experiment can be said to be accurate. The goal of a good experimentalist is to then evaluate the experiment to try to find the places where the uncertainty can be reduced. If the methods are valid, then the true value should lie within the new, smaller range determined by the uncertainty. This would give an even more accurate value for the quantity being measured.

Identify the largest contributor to uncertainty

You can use your percentage uncertainties to evaluate the experimental techniques. You should try to reduce the factor which has the biggest contribution to the overall uncertainty. Remember that if the equation involves powers, e.g. T^2, then the quantity with the largest percentage uncertainty may not necessarily contribute the most to the final uncertainty.

Techniques for reducing uncertainty

You need to be familiar with various methods to reduce uncertainty and increase the accuracy of measurements, such as the use of a set square to check alignment. Many techniques have a common idea behind their application, so that, once understood, they can be applied in many different circumstances. Essentially they reduce the percentage uncertainty by increasing the value of the mean while maintaining the absolute uncertainty.

When measuring multiple objects stacked or lined up, the total length is increased, while the absolute uncertainty remains fixed. Using n objects or n oscillations the percentage uncertainty is reduced by a factor of n.

Object	Typical measurement	Absolute uncertainty	% uncertainty	% uncertainty using 10 objects
Coin	Thickness of 1 coin: 1.5 mm	±0.01 mm using micrometer	$\% U = \dfrac{0.01}{1.5} \times 100 = 0.7\%$	$\% U = \dfrac{0.01}{15} \times 100 = 0.07\%$
Paper clip	Length of 1 paper clip: 50 mm	±1 mm using ruler	$\% U = \dfrac{1}{50} \times 100 = 2\%$	$\% U = \dfrac{1}{500} \times 100 = 0.2\%$
Time period of oscillating object	Time period for 1 oscillation: 1 s	0.1 s due to experimenter's reaction time	$\% U = \dfrac{0.1}{1} \times 100 = 10\%$	$\% U = \dfrac{0.1}{10} \times 100 = 1\%$

Worked example 4.13

Combining uncertainties

A student is investigating the relationship between force and change of momentum. A load is attached to the end of a string, fed over a pulley and attached to a glider on an air track. For a certain reading, the the load consists of four similar objects. The mass of one of these objects is found to be 2.200 kg. The student claims that it does not make any difference if they measure the mass of one object and multiply by four, or find the mass of all four objects together on the balance.

Determine the percentage uncertainty in the force and comment on the students claim. **(5)**

force = weight of the objects = $m \times g$

$g = 9.81 \text{ m s}^{-2}$, which means the absolute uncertainty is 0.01 m s^{-2}.

The value should be quoted as $g = 9.81 \pm 0.01 \text{ m s}^{-2}$.

percentage uncertainty $= \dfrac{0.01}{9.81} \times 100 = 0.1\%$ ✔

This uncertainty is the same for either method.

The mass is quoted as 2.200 kg.

absolute uncertainty = ±0.5 × smallest scale reading ±0.0005 kg

Four masses separately:

With four masses, total uncertainty (add the absolute uncertainties) = 4 0.0005 kg = ±0.002 kg (±2 g)

percentage uncertainty in total mass (when measured separately) = $\dfrac{0.002}{8.8} \times 100 = \pm 0.02\%$

total percentage uncertainty in force = 0.1% + 0.02% = ±0.12% ✓

Mass measured together:

percentage uncertainty = $\dfrac{0.0005}{8.8} \times 100 = \pm 0.006\%$

total percentage uncertainty in force = 0.1% + 0.006% = ±0.106% ✓

The student is incorrect. The percentage uncertainty in the mass is four times greater when the mass is determined for each object separately. ✓

percentage difference = $\dfrac{\text{difference}}{\text{mean}} \times 100 = \dfrac{0.12 - 0.106}{(0.12 + 0.106)/2} \times 100 = 12\%$

The overall percentage uncertainty in the force is 12% larger when the masses of the objects are measured separately. ✓

Apply scientific knowledge to practical contexts

Although there will be questions based on the core practicals, there will also be many questions which use theory from a different part of the course and ask you to explore how this applies in a certain practical context. The key here is to consider the underlying principles behind the process. You need to consider what physics you know about this situation.

Example: heat transfer

Using a temperature sensor to measure the temperature change of water heated by an electric heater, you find that the temperature of the water keeps rising even after the heater is switched off. How can this be? You cannot just say that the heater is still hot so the water keeps getting hotter, and there is not a substantial lag between the temperature sensor and the water temperature. To answer questions such as these, you need to think about the underlying principles behind the process. This is really about energy transfer in the form of heat. Heat energy always moves from hotter objects to cooler ones. There is no temperature change when two objects are in thermal equilibrium. So why might the water keep on getting hotter even though the heater has been switched off? The two objects cannot be in thermal equilibrium yet, so the heater must have been hotter than the surrounding water even though it had been switched off. The peak water temperature shown by the temperature sensor marks the point where the heater and water reach thermal equilibrium.

Evaluate the design of an unfamiliar instrument

In addition to knowing the appropriate techniques to use with standard laboratory equipment, such as vernier callipers, oscilloscopes and light gates, you also need to be able to evaluate the design of equipment which you may never have seen or heard of before.

You should ask yourself how an unfamiliar instrument compares with an instrument you have previously used, or that has previously been described. The term 'evaluate' also means that you need to assess the negative implications of using the new device. You need to look analytically at this instrument and ask a series of questions about it in order to state what features it has which enable it to be better (or worse) than the alternative measuring device and then explain *why* this feature makes it better or worse.

Worked example: 4.14

Evaluating a new piece of equipment

Your teacher presents you with a new type of compass to investigate magnetic fields. Discuss the steps you could take to evaluate its accuracy. (3)

Step 1: Is there a zero error?

A key feature of good experimental design is to predict and then suggest ways to eliminate sources of error which will produce systematic errors. The most obvious example of this is a zero error, i.e. does your instrument give a false reading when the true value should be zero?

Step 2: Does it have improved resolution?

First determine whether it is a analogue or a digital device. For a digital device, does it have an extra decimal place of measurement? It is possible to **interpolate** to half the scale markings on most analogue scales. If the spacing between the markings is increased, you may be able to interpolate to plus or minus one tenth of the scale. This would give the instrument a **higher resolution** and make it more **sensitive**.

When assessing experimental design, you need to consider how the experiment can or should be set up. Exactly how will the experiment ensure that, for example, the screen to show the interference is parallel to the double slits. You need to be able to explain clearly a sensible, repeatable way of ensuring the apparatus is arranged accurately, generally using the apparatus that is available during your required practical activities. Occasionally you may be asked to suggest how you may use a new, unfamiliar piece of apparatus, but in general examiners will be looking for you to show them that you know what equipment is generally available in a science lab and that you know how it can be used.

Step 3: Can it be used in the context of the particular investigation?

When assessing experimental design, you need to consider how the experiment can or should be set up. Will it actually be possible to use it in the particular situation, i.e. will it interfere with the experiment due to its shape or size?

> ✓ **Exam tip**
>
> **Evaluating experimental techniques and calculations**
>
> When asked to evaluate experimental calculations, state an assumption that has been made in the calculations.
>
> For example, when a system uses multiple different springs, does the student assume that each of the springs has the spring constant (stiffness). Are the masses the same, is the ruler stiffness the same, has the student taken into account the masses of all of the different components, e.g the nuts and bolts holding an object onto a balance point?

Step 4: Does the device reduce a particular source of error in the method?

Is there a reflective material alongside the measuring scale that you can use to align your eye with the pointer to remove the parallax error?

Does the new scale give a direct reading rather than having to use one reading to perform a calculation or second reading?

> To check for a zero error, remove the instrument from the experiment before starting the investigation and observe the reading on the compass. Since the device does not have a 'zero' value, you need to compare the reading with other instruments, preferably already calibrated (zeroed) devices. If the device gives the same reading, then there is no zero error. ✓
>
> The compass is an analogue device, so a higher-resolution scale will enable a more accurate reading, e.g. 1° scale divisions. ✓
>
> If the instrument has a magnifying lens to make the position of the pointer easier to identify, it will enable a more accurate reading. ✓
>
> Also, if the pointer is longer, it will increase the sensitivity because a smaller change in the input will produce a larger movement in the tip of the pointer, again increasing the accuracy. ✓
>
> If there is reflective material behind the pointer, the parallax error will be reduced, enabling a more accurate reading. ✓   Max 3

Evaluate the effect of changing an aspect of the procedure or equipment

You need to be able to predict and explain the effect of varying different aspects of the experiment.

These types of question can be called 'what happens if...?' questions. Here are some examples:

1 **Explain the effect of changing the time base setting on an oscilloscope from $0.5\,\text{m s}^{-1}/\text{div}$ to $0.2\,\text{m s}^{-1}/\text{div}$. (2)**

This change has made the oscilloscope more sensitive on the *x*-axis. What would have fitted in 1 square will now take 2.5 squares. This could have a negative effect, in that the wave trace can no longer fit onto the screen, so that measurement of the time period is more difficult, but it may have a positive effect by increasing the length that you are using to calculate the time period. The absolute uncertainty remains the same, i.e. half a square, but since the overall wave period is 2.5 times longer, the percentage uncertainty is reduced by a factor of 2.5 times.

2 **Explain what would happen to the first harmonic frequencies produced on a stretched string if the tension is increased to a very high value.   (2)**

Since $f = \frac{1}{2l}\sqrt{\frac{T}{\mu}}$ it is tempting to jump in and just say that the frequencies increase as the tension increases. But does anything else change as the tension is increased to a very large value? The string itself may change shape. As it gets stretched, it will get thinner, i.e. its diameter would decrease. This means there is less mass per unit length, i.e. μ decreases rather than being a constant, as could be the case at low tension. The effect of a decreasing μ would be to increase the frequency to an even higher value than would have been predicted with μ being constant.

The difference between...

A student is investigating an oscillating system. The total length of the oscillating object is made up of two lengths l and l_0.

During the experiment l_0 is fixed, l is systematically increased and the time period, T, is measured each time.

The student plots a graph of $\sqrt{l + l_0} - \sqrt{l}$ against $\frac{1}{T}$, which produces a straight line with a positive gradient.

A different student did not line the ruler up vertically with the string when they took their reading for l_0. Describe how this student could have improved their experimental technique and explain the effect of this error on the graph. (3)

Response (a)	Response (b)
The student should have used a set square placed against the ruler and the bench to ensure the ruler was vertical ✓. This would ensure that the ruler was straight. The error is systematic because it is the same each time. This makes all of the points bigger by the same amount so the gradient stays the same, but the line is higher and the y-intercept is higher.	The student could have ensured that the ruler was vertical by using two set squares arranged against the ruler perpendicular to each other, but both in the vertical plane. ✓ If the ruler was not straight the student will have overestimated the length measurement (the reading would be the hypotenuse of the triangle). Each value of $\sqrt{l + l_0} - \sqrt{l}$ will be bigger than it should be, but not by a constant amount. As l increases the systematic error will make a smaller and smaller difference, so the y-value on the graph increases ✓ at a decreasing rate. The student will have a curve of decreasing gradient instead of a straight line. ✓ The systematic error on the y-axis also means that there will be a non-zero y-intercept. (✓)
Student (a) has correctly suggested a way to improve the experiment, and although they have stated that the line would be higher they do not get credit because of insufficient reasoning. They have made a common mistake — thinking that a systematic error always increases values by the same amount so the gradient stays the same.	*Student (b) has given a more detailed suggestion for improvement and also correctly analysed the equation to see that although all of the points for $\sqrt{l + l_0} - \sqrt{l}$ are bigger, they are not all bigger by the same amount. Student (b) has also correctly stated that the new graph will not pass through the origin.*

Worked example 4.15
Capacitor discharge

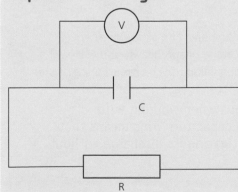

Figure 4.3 Circuit diagram showing a capacitor, C, discharging through a fixed resistor, R

The capacitor is discharging through a fixed resistor, R, and the pd is measured using a digital voltmeter.

(a) Describe how the circuit shown in Figure 4.3 can be used to determine the resistance, *R*, of the fixed resistor. **(3)**

(b) Deduce how this value would change if you carried out the experiment with an analogue meter with the same resistance, *R*, as the fixed resistor, instead of the digital voltmeter. **(3)**

(a) You would record the voltmeter readings over time. (The interval of the timing will be set after a preliminary test to see how quickly the capacitor discharges — if the discharge is quick, around 1 minute, then the interval should be 5 s; if the discharge is much longer, then you could use an interval of 10 s). ✓

Plot a graph of ln *V* against *t*. ✓

You would compare the capacitor decay equation to the equation for a straight line in order to determine the resistance, $R = \dfrac{1}{C \times \text{gradient}}$. ✓

This technique is covered in Chapter 2.

(b) The experiment would now look like Figure 4.4.

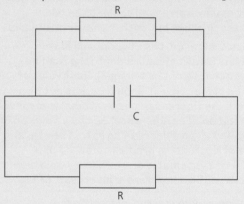

Figure 4.4 The voltmeter now acts like a resistor connected in parallel with R.

Use the resistors in parallel equation:

$$\frac{1}{R_t} = \frac{1}{R_1} + \frac{1}{R_2}$$

Therefore:

$$\frac{1}{R_t} = \frac{1}{R} + \frac{1}{R}$$

$$\frac{1}{R_t} = \frac{2}{R}$$

Therefore:

$$R_t = \frac{R}{2} \checkmark$$

In other words the resistance of the circuit is halved. This would also halve the time constant because the capacitor would discharge in less time. The gradient of the graph drawn above would be twice as steep. ✓ So the experimenter would not get the correct value for the resistance of the fixed resistor — its value would be half of the correct value.

The higher resistance of the digital voltmeter also has the bonus when working with capacitors that it prevents significant charge leakage from the capacitors. This means that you can wait longer between charging the capacitor and investigating its discharge without affecting the results.

Application to the exam

A student is investigating the variation in intensity of gamma radiation with distance. The student uses a cobalt-60 source source with a Geiger–Müller (GM) tube connected to a counter a distance d away. The count is recorded for 2 minutes for values of d from 50 mm to 950 mm, with an interval of 100 mm.

(a) Explain how the student could collect and process data from the counter to show that the intensity of gamma radiation follows an inverse square law. (5)

(b) Another student says that the method is flawed because they measured the count for the same amount of time for each distance.

   Discuss this argument. (5)

(c) It is suggested that the experiment could be improved in two ways by turning the GM tube sideways on to the radiation:

   1 The experiment would be able to work with sources that emitted alpha and/or beta radiation in addition to gamma radiation.

   2 The accuracy of the experiment would be improved.

Evaluate the effect if the student turned the GM tube sideways on to the radiation. (4)

B-grade answer

(a) They would find the background count rate for 2 minutes and do this three times and calculate the mean. They would then measure the count rate at each distance three times and also calculate a mean. ✓ They would then calculate the corrected count rate by subtracting the background count rate from the count rate. ✓ They would then use a graph of $\dfrac{1}{\sqrt{C}}$ against distance ✓ to show the inverse square relationship.

The candidate has confused the count rate with the count — the count rate is the $\dfrac{\text{count}}{\text{time}}$. They state marking point 3 and 4 in this first section. They then correctly state the graph to be plotted but fail to explain the significance of the gradient (or intercept) in verifying the inverse square law. **Marks awarded: 3**

(b) The uncertainty in the count rate, C is proportional to $\sqrt{C}$. The percentage uncertainty in C is given by: $\dfrac{\sqrt{C}}{C} \times 100$.

As d increases C decreases. This means that the percentage uncertainty increases at larger values of d. To reduce this increased uncertainty they need to increase C, which they can do by increasing the time they measure the count for at larger distances. ✓ Therefore I agree that the original experiment was flawed because they used the same time for each distance. They should have measured the count for a longer time period at larger distances. ✓

The candidate has again confused count and count rate. But they have identified the correct underlying principle, $C = \pm\sqrt{C}$. They have then followed this reasoning through and have arrived at a sensible conclusion. However, they have not used the data in the question to quantify their argument. **Marks awarded: 3**

(c) There is now a larger area for detection of radiation so the count rate will be increased. ✓ This means the count rate will be more accurate, since $C = \pm\sqrt{C}$. Also, in the perpendicular arrangement the uncertainty in the distance will be decreased. This is because in the old arrangement the radiation could be detected at any distance within the GM tube, whereas in the new orientation the uncertainty in the place where the detection occurs is reduced. ✓

The candidate gives a detailed account of two ways in which the accuracy of the experiment would be improved (although they continue to confuse the term count and count rate). However, the candidate fails to address the first point about improvements with different types of source. **Marks awarded: 2**

A/A*-grade answer

(a) Before removing the source of the container, the student should measure the background count. Because the count will be low they should record the count for a longer period of time than the 2 minutes used with the source at different distances, e.g. at least 5 minutes. ✓ This is because the uncertainty in the count, $C = \pm\sqrt{C}$. They should then divide the total count by the time to get the background count rate. ✓

They will then record the count for 2 minutes at each distance, d, repeating this twice and calculating a mean. They will divide the mean count by 2 to get the count rate (per minute). They will then calculate the corrected count rate, C, using the equation: corrected count rate,

$$C = \text{count rate} - \text{background count rate} ✓.$$

They will then plot a graph of $\dfrac{1}{\sqrt{C}}$ against d. ✓ The graph will be a straight line which does not pass through the origin if the count rate for the gamma source is proportional to $\dfrac{1}{d^2}$, i.e. the count rate follows an inverse square law. ✓

This is a very detailed answer, which address all of the key points involved in both the collection of data and the processing of the data, along with the subsequent analysis of the graph. **Marks awarded: 5**

(b) Since radioactive decay is a random process there will always be a random error in the count rate. ✓ The uncertainty in $C = \pm\sqrt{C}$. ✓ The percentage uncertainty in $C\left(\dfrac{\sqrt{C}}{C} \times 100\right)$ increases when C decreases. This means that as the distance, d, increases, the count decreases (following an inverse square law), so the percentage uncertainty increases. ✓ In this particular experiment the range of d is from 50 mm to 950 mm. Therefore the final reading is $\dfrac{950}{50} = 19$ times further away, so the count rate will be $19^2 = 361$ times smaller. The percentage uncertainty in $C = \dfrac{\frac{\sqrt{C}}{\sqrt{361}}}{\frac{C}{361}} \times 100 = \dfrac{\sqrt{\frac{C}{19}}}{\frac{C}{361}} = \dfrac{\sqrt{C}.361}{C.19}$ $= 19\sqrt{\dfrac{C}{C}}$, i.e. 19 times greater than at 50 mm. ✓ Therefore to reduce the percentage uncertainty at larger values of d, they should increase the time the count is recorded for ✓ (up to 19 times greater than at the beginning — although they could record for 1 minute at the shortest distances, otherwise the experiment will take too long).

An excellent response. The candidate has carried out a numerical analysis based on ideas in the question and data provided in the question stem and come to a reasoned judgement at the end.
Marks awarded: 5

(c) Alpha particles have a very small range in air (3–5 cm) so they will not reach the detector inside the GM tube in either orientation. ✓ However, since the range of beta particles in air is 15 cm, in the original orientation both beta and gamma radiation could reach the detector for the shortest distance measurements. Turning the GM tube perpendicular to the source means that the beta particles can no longer reach the detector because they cannot penetrate the sides of the tube. This means that the readings will only be due to gamma radiation, even if the source emits alpha, beta and gamma radiation. ✓

In the original orientation you measure a distance from the window of the GM tube to the front of the radioactive source. The source is actually on a foil which lies a few millimetres behind the mesh of the cup source holder; this issue is not changed by the new experimental arrangement. The average detection point of the radiation is somewhere within the GM tube, so in the perpendicular orientation the uncertainty in the place where detection occurs is reduced ✓, and the distance measurement, d, will be more accurate ✓.

Therefore, turning the GM tube sideways on to the radiation improves the experiment in two ways.

Both parts of the question are addressed in detail, but both paragraphs could have been written more concisely. Parts of the first two sentences of the second paragraph do not contribute any marking points to the question being asked. Remember to focus on the question being asked. Writing superfluous information wastes time that could be spent on other parts of the exam.
Marks awarded: 4

You should know

> Practical skills will be assessed directly for the practical endorsement through the CPAC and indirectly assessed in your written exams.
> The key terminology related to practical physics must be learned.
> Planning an investigation involves defining a problem, identifying the variables, obtaining sufficient data and using appropriate equipment and techniques.
> Analysis of the data in results tables and graphical plots can be used to identify relationships between the variables in an investigation.
> Use uncertainties in measurements and calculated values to evaluate investigations and suggest improvements to experimental procedures.
> Unfamiliar equipment is evaluated by carefully examining the purpose and processes needed to make measurements with it.

5 Study skills

Introduction

To be successful in linear exams, which are assessed at the end of a 2-year course of study, it is essential that you review the material throughout the course. This process of looking back at content you have already covered and making notes to improve your understanding is called **revision**.

Effective revision will look different for different people. You each have a different starting point and different outside influences, and may prefer different revision techniques. That said, there are certain fundamental and important features that must be in place if your revision is to be effective. We will look at the neuroscience of revision and use this to develop strategies to improve your ability to remember, apply and evaluate in exam situations.

The neuroscience of memory formation

Short-term memory typically lasts only around 30 seconds, so any information that can be recalled after that time is a long-term memory. It is vital, then, that you move information from short-term to long-term memory. When a memory is stored in your long-term memory, neurones make new physical connections (called synapses) between each other. All strategies to maximise learning and long-term retention must be based around effective creation of these new neural connections.

There are two main factors to consider: (1) convincing your brain that the information is important enough to transfer into long-term storage, and (2) not over-working the brain, which prevents new memories being formed.

The rest of the chapter goes into more detail about effective revision strategies and looks at how and why they work, but in essence there are two key messages about effective revision:

1. **You remember what you think about.**
2. **Take regular breaks.**

Core study skills

The key skills for effective revision

Put yourself in control: organise your time

Organising your time effectively will maximise your ability to help your brain create new memories without being overloaded and then strengthen those memories. The following key features should be part of your revision schedule.

Focus

The key to learning new information and retaining it is focus. By directing your attention onto something and by thinking about the concept, you will tell your brain that this information is important and needs to be stored as a new long-term memory.

Many neuroscientists think that playing music whilst you are revising will have a detrimental effect on your ability to remember the information you are studying because some of your brain's attention is directed towards the music, and away from your work. If background music has a negative effect on your focus, you definitely should not revise with one eye on social media or Netflix!

Take breaks

As your brain is busy storing different patterns of information, it can get confused and overloaded. The best way to prevent this overload is by taking regular breaks. This will help maintain your focus.

A popular schedule for combining work and breaks is to work for 25 minutes and then break for 5 minutes. In the break you should get up and do something else, e.g. stretch, walk around or make a drink. After four such sessions you should take a slightly longer break, e.g. 15 minutes. A simple way to keep track of the sessions is to use a timer on your phone, but you can also download specific apps that help you track how much work you are doing for each subject — as a physicist you may find the data interesting and motivational!

Space out your sessions: don't cram

You should space out your revision sessions, rather than cramming them all together on one day. If you plan to do 5 hours of revision on particle physics, then doing 1 hour per day for 5 days is far more effective for long-term recall than doing 5 hours on one day. Despite knowing this, we often fall into the trap of cramming. There are two main reasons for this: first, and most obviously, we did not start our revision early enough and we feel the need to rush through it — see the next point for remedying this. The second main reason is that it actually feels easier for our brains to cram. The subsequent hours of study feel easier when the previous information gives us a sense of familiarity with the material. Unfortunately, this is counter to all current thinking on making learning effective. Learning should feel hard, because you learn by recalling information from memory rather than having a sense of recognition about the content. There will be more on this in the next section.

Revision timetable

Making a revision timetable is an important feature of your revision because it enables you to organise your time and work effectively. Your plan can include sufficient time for shorter, more effective, sessions with breaks, and because you do not have to worry about running out of time you can focus on exactly what you need to be doing at that moment, rather than worrying that you should be revising something else. The revision timetable puts you in control. This makes you more relaxed, which is the ideal state for making and strengthening memories.

The fundamentals of effective revision technique

The previous section talked about getting the environment right and setting up the structure to focus on what you are revising. Now we can look at the strategies you should use within those sessions. This section will look at remembering key information and understanding it. The next section will look at how to build the higher-order skills of applying, evaluating and creating into your revision.

Process the information

All effective revision techniques have one thing in common. They make you *process and think about the information* in some way. This means that you should not just read over your notes, or read over exam questions or mark schemes. You need to interact with the information and process it in some way, summarising it, transforming it into a diagram, answering an exam question etc. This is similar to the way we discussed active and passive reading in Chapter 2 — you cannot just read over the information and hope to have enough recall to be useful in an examination. The more thinking you do as part of the processing the better, because the more you think about a particular concept the more you indicate to your brain that the information must be important, and therefore the more neural connections are devoted to understanding and remembering the associated information.

Test yourself

Many (often more passive) revision techniques involve recognising information, rather than recalling it. Often students feel bored, or feel that they have fully revised when they look over their notes or look at a practice paper question and feel a sense of familiarity. You are tricked into thinking that you understand it, but there is a big difference in an exam situation between being familiar with a topic and being able to recall information about it accurately. It is easier to use techniques promoting recognition: reading through notes, scanning mark schemes etc., but *effective revision makes you think*.

If it does not feel challenging, your revision is not working.

You must test yourself to ensure that you can recall the information accurately (or know where to find it on the data sheet).

With these important criteria in place, let us look at some of the most effective ways to remember information for your physics exams.

Remembering information

In previous chapters we have discussed the importance of learning specific information from the specification, such as definitions, laws and units. Here we look at some of the best ways to help you learn the information.

Flashcards

When something needs to be learned exactly, one of the best learning techniques is to use **flashcards**. Flashcards are essentially a way to test yourself on information by writing a question or key term on one side, and writing the answer or definition on the other. There are lots of free electronic flashcard apps and websites (it is often easier to enter your information via the website using a keyboard, and then sync your flashcard sets to your mobile device). The electronic format is good because you always have the information on you and can test yourself whenever you have a spare minute or two, even if you are just waiting for the bus.

For example:

Key term	Definition
Principle of moments	The **sum** of the clockwise moments = the **sum** of the anticlockwise moments.
Material dispersion/ spectral dispersion	The speed of light in an optical fibre depends on the wavelength of the light (red travels faster than violet in glass), so if white light is used the pulse become longer.

When using the flashcards you should always have the key term face up and test your recall of the definition (doing the opposite simply encourages recognition, which will not be enough in the exam).

You can also make flashcards for specific examination question mark schemes.

Activity

Spaced repetition is a revision technique based on the idea that you should repeatedly test yourself on information, but it also says that you should *test yourself more frequently on the bits that you find the hardest to remember*. At its simplest you could create five 'boxes' to store flashcards with different review frequencies depending on how easily you can recall the information:

1 Every day — this is information that you are really struggling to learn

2 Every other day

3 Once per week

4 Every other week

5 Just before the test — this is material that you have excellent recall of

Make a set of flashcards for the topic you are studying or have already studied and try to develop your own spaced repetition system to supercharge your memory.

Memory palace/method of loci

The most commonly used memory technique among memory contest champions is the 'memory palace'. This is also known as the method of loci. The technique, which was known to the ancient Greeks and Romans, is ideally suited to remembering information that follows sequentially, such as speeches, but also works for faces, digits, lists of words etc. It has been used to recall pi to over 65 000 digits.

How does it work? You pick a place that you know very well, such as your house. You then imagine objects at specific locations around your house, e.g:

→ on your drive, at your porch door, at your front door
→ entrance hall
→ lounge:
- sofa
- coffee table
- shelves
- fireplace
- television
- window ledge

The technique works more efficiently if you always take the same path around the house, e.g. clockwise, and it is even better if you have the same number of locations in each room, e.g. five. Then you know if you have got all of the places. To make the memories stronger you should try to make the images more vivid by making an emotional connection to what is happening, e.g. by making it funny.

Worked example 5.1

Defining simple harmonic motion

Simple harmonic motion is defined as oscillating motion in which the acceleration is proportional to the displacement and always in the opposite direction from the displacement.

Using your lounge as the room with the objects being the television, the light fitting, the sofa, the coffee table and the fireplace, you could produce your own memories within the palace:

Lounge

	Location	Event	The physics
1	Television	Your physics teacher is on 'Britain's Got Talent' singing a (terrible) simple harmony and swaying with the microphone.	Simple harmonic motion
2	Light fitting	The light fitting is swinging from side to side.	Motion with oscillations
3	Sofa	You are sitting on the sofa and can hear a car speeding off outside — the driver must be flooring the accelerator from a standing start right on your street.	Where the acceleration
4	Coffee table	The coffee table is *propped* up at one side (making it higher on that side) so all of the cups keep sliding along the table (displacing).	Proportional to the displacement
5	Fireplace	The flames in the fireplace are upside down. They are coming down out of the top of the fireplace — they always do that in this strange fireplace.	Always in the opposite direction (to the displacement)

The events are supposed to be strange — there is a lot of evidence to suggest that the stranger the event is, the more likely you are to remember it. You should be able to find many other areas where the memory palace technique will be able to help you remember key information.

Activity 5.1

Build your own memory palace. Think of the main objects in your room, then move to the next room and repeat this throughout your house. It will take some time to 'build' your palace, but then you can start to use it to remember anything you want.

Use it to remember the prefixes: tera ($\times 10^{12}$) — a terrifying (tera-fying) monster with 12 arms is waiting at your front door. (Tera derives from the Greek word meaning monster.)

Try to use your memory palace for the derivation of $pV = \frac{1}{3} Nm(c_{rms})^2$.

Take it further

Joshua Foer visited the US Memory Championships to report on the contestants and their amazing abilities. Through his interviews he learned that all of the contestants used the same technique: the 'memory palace'. In the TED talk 'Feats of memory anyone can do', Foer describes how he developed his ability to use the memory palace technique and subsequently used this to win the US Memory Championships. His TED talk will inspire you to make the most of the amazing memory capabilities that you have: https://www.ted.com/talks/joshua_foer_feats_of_memory_anyone_can_do

Understanding

The previous sections have talked about remembering information. As we have mentioned, having knowledge of, say, a definition is necessary, but you cannot succeed on a physics course by memorisation alone. The key to success is not just knowledge, but also understanding. Obviously, knowledge of the concepts and terms involved is an important prerequisite for being able to apply physics principles and ideas to different situations and to evaluate new information.

Use the specification as a checklist

With physics, if you truly understand a concept, as determined by your ability to answer several past examination questions on the topic correctly, then you have less need of memorisation than in many other subjects. The symbol equations and values for constants, along with their units, are given to you in the exam on the data sheet. You just need to know what the symbols stand for.

For example:

$$g = \frac{-\Delta V}{\Delta r}$$

You need to know that:

➜ g is the gravitational field strength at a point a distance r from the centre of a planet.

→ $\frac{\Delta V}{\Delta r}$ is the gradient of the potential curve at a distance r from the centre of a planet.

Read around the subject

Your understanding can be improved by increasing your background knowledge (as discussed in Chapter 2). Background knowledge, also called prior knowledge, is essential in making connections between the new information and the experiences and ideas that we already know. It is the opposite of Homer Simpson's view: 'Every time I learn something new, a little of the old gets pushed out of my brain.' In fact: *the more you know, the easier it is to learn more.*

So how can you increase your store of prior knowledge?

Video

Subscribe to physics online video channels.

There are so many excellent physics videos online. Obviously you have to check that the information they present is scientifically accurate. A few of the best (entertaining and educational) are:

→ Sixty Symbols www.sixtysymbols.com/
→ MinutePhysics www.youtube.com/user/minutephysics
→ Veritasium https://tinyurl.com/mavy9ns
→ Vsauce https://tinyurl.com/mbytvak
→ CrashCourse — Physics Playlist https://tinyurl.com/jszkbwn
→ SmarterEveryDay www.youtube.com/user/destinws2

You should also keep your eye on BBC documentaries on physics (on iPlayer, click: Categories → Science & Nature).

Audio

Subscribe to physics podcasts.

Use your podcasting app of choice to listen to the latest information about physics and the universe whilst you are travelling or do not have the opportunity to look at a screen. Some good places to start are:

→ The Infinite Monkey Cage (BBC)
→ Science Weekly www.theguardian.com
→ Little Atoms
→ The Naked Scientists Podcast (University of Cambridge — BBC)
→ The Skeptic's Guide to the Universe
→ Science Talk
→ The Reith Lectures (BBC)

Use a range of resources

There is more than one way to present an idea in physics, so if you have not fully understood it after first meeting it in class and consulting your textbook, try a different resource to get an alternative description or explanation.

Here are some good online resources:

→ Antonine Physics is a good alternative set of notes with some worked examples www.antonine-education.co.uk

→ www.topphysicsgrades.com is my physics website. It has condensed revision notes, explainer videos and example questions.

→ www.physicsandmathstutor.com has brief revision notes and lots of past paper questions.

→ The DrPhysicsA Youtube channel has whiteboard style explainer videos which cover the main A-level topics. It also goes beyond A-level with videos such as 'Spin Orbit Coupling' and 'The Strength of the Nuclear Force' www.youtube.com/user/DrPhysicsA

Some of the most successful students are part of a motivated and supported informal study group. Study groups give you the opportunity to ask questions of other students who have a better grasp of particular concepts. Perhaps they can see how to apply the theory in the context of a particular question you are struggling with. It also gives you the opportunity to help others. This is one of the most effective ways to improve your understanding and long-term retention. If you can successfully teach another student about an area of physics, then you definitely understand it, and the bonus is that because the act of teaching others forces you to think about the concepts involved, you will create longer-term memories.

You should also ask your teacher for help if you are struggling with a particular topic or question. Your teachers appreciate proactivity on your part, so if you seek their help well in advance of any deadlines, they will be very glad to give you as much help as you need.

Higher-order study skills

Applying

Application of knowledge and understanding questions account for the largest proportion of marks in your A-level. Too many students spend their time making revision notes, perhaps even testing recall of this information, but they do not prioritise the practice of *applying* their knowledge and understanding.

A sporting analogy works well. You could practise set plays without the opposition and you would be able to do them without a problem, but the only way to get better in a game situation is to play the game. You will find that the formations and strategies you planned to implement need to change slightly depending on the opposition or the environmental conditions on the day. This is the same with school examinations. You have to adapt your performance to the context of the question. The absolute best way of practising applying your knowledge and understanding is by answering past examination questions, so ensure they are integrated into your revision from the beginning.

As well as answering full papers, you can ask your teacher to prepare booklets of past paper questions on the sections of the specification that you have covered. This makes your revision much more effective: you can learn the theory, definitions etc., and can then assess your level of understanding against exam-style questions.

Evaluating

The skills of analysing, interpreting and evaluating account for 25% of the marks available in your exams. This book is full of guidance on how to develop your evaluation skills in the contexts of tables of data, reading sources and practical procedures, but you can also develop your evaluating skills whilst revising.

Mark schemes: formative assessment

When you mark your written responses to exam questions against the mark scheme criteria, you are evaluating your answers. When completing questions there are essentially two types of assessment you can carry out: summative and formative. Summative assessment is essentially about grading. You evaluate your response to give it a score, a number of marks which tells you how well you did compared with the mark scheme standard. To be effective, though, your revision needs to be focused on formative assessment. Formative assessment is where you evaluate your answers to identify areas where you can focus subsequent revision. As part of this higher-level evaluation you can find two main areas where you can improve:

→ Subject knowledge — do you need to add a particular key term or phrase to your flashcards? Do you need to review your understanding of gravitational potential and then practise some further questions on this topic?

→ Exam technique — should you have used data from the graph to support your answer? Why did you only include two key marking points instead of adjusting your answer to the 3 marks allocated for the question?

Insights from your formative assessment can then be included into your revision notes. Formation assessment can be gleaned from a number of sources.

Examiners' reports: key insights from the examining team

The examining team carries out its own evaluation of the candidate responses and summarises the findings in the examiners' report for each exam. The reports comment on general patterns in candidate responses and highlight areas of particular strength or weakness. You should take a special note of any areas where the examiner has commented that improvements need to be made, as future exams will contain questions designed to assess whether candidates have improved in these areas.

As a candidate aiming for an A/A* grade, you should consult the examiners' reports for all previous past papers and add these insights into the relevant area of your revision notes.

Creating

Revision and independent learning activities that promote the creation of original material or generate links between information will be highly effective. Creating is a higher-order skill that really makes you think about the concepts that you are trying to learn or revise, which as we have seen means that you will retain the information for longer.

Activity 5.2

Even if you have not seen the benefit of mind maps in the past, make a mind map of the particle section of your specification. Aim to include: classification of particles, particle interactions, particles and anti-particles, and conservation laws. If you have space, you may even be able to include constituents of the atom and explanations of the strong and weak forces and Feynman diagrams. As you use mind maps more frequently, your skills will develop and you will find them even more effective, both for remembering and deeper understanding.

Mind maps

Mind maps are an excellent way, not only of learning factual information, but also of seeing and making connections between different areas of physics. A specification area such as particle physics is ideally suited to mind mapping. A mind map will clearly show the classification of fundamental particles along with the connections to the conservation rules.

By selecting the information yourself and personalising the content, you are thinking about the concepts involved and helping long-term recall. You should be able to close your eyes and walk around your mind map. The areas where you get stuck are the areas you need to focus on. Aim for 100% recall. Mind maps are also excellent for revision because new information can be added in as your revision progresses. For example, notes from examiners' reports can easily be added as you complete past papers.

Comparison tables

Creating a comparison table is a great way of seeing the connections between two or three different concepts. As a tool it is well suited to directly comparing and contrasting between a small number of ideas, or comparing different items against a particular list of features. In fact this is where you tend to see comparison tables, e.g a website comparing different cars or mobile phones. In your revision you can use it to see similarities and differences between closely related topics such as different types of force, types of radiation, or types of field.

Model answers

Creating your own set of model answers for different styles of question is a really effective revision technique. It practises many different skills, including: recall, application of knowledge and understanding, evaluation of mark schemes and examiners' report comments, and then finally creating your own content blending together all of these other resources. Creating model answers for extended writing questions is particularly useful as it also requires you to practise analysis, drawing conclusions and making judgements, all skills of a highly competent physicist.

As part of your informal study group you could even have a go at writing your own examination-style questions and mark schemes. Allocate each of you 10 marks to assess a different area of the specification. Review the past paper questions on the same topic and then create your own question. You will find this activity challenging. Thinking is difficult, but thinking is key to learning and long-term retention of information.

Analysing

We have already discussed a kind of 'gap analysis' that you can do as part of your formative assessment when you are marking your work. But how can you further develop your analysis skills as part of your revision?

One way is to analyse past papers to identify which of the questions are designed to test which assessment objective. Remember that some longer questions may assess more than one objective. You should be able to easily identify AO1 objectives and answer these questions using mostly recall. Do you have the required information

Activity 5.3

Create a comparison table for fields: gravitational, electrostatic and magnetic. Think about the criteria that would be useful to compare, e.g. source, range, strength, direction of field lines...

Activity 5.4

Arrange for you and some of your friends from the physics course to write some exam questions (and mark schemes) to test each other. You could even get your teacher to photocopy them for you into an exam booklet. Each of you should answer the full paper (you should do well on your own question!) in exam conditions and then mark it according to your mark scheme. Any discussion generated about alternative acceptable answers is an excellent learning opportunity. This kind of debate occurs after each real exam when the examiner team gets together to decide upon the final mark scheme.

in your revision notes? Where are the AO3 questions? What are the command words associated with AO3, and how would you structure your answers for these questions?

You can also analyse the specification content of each past paper. Are there patterns between papers. Are some concepts assessed each year? Is there anything that hasn't yet been assessed? Remember that every mathematical skill and piece of content from the specification must be assessed over the lifetime of the specification. As you get nearer to the end of the specification's lifetime, it may be possible for you to try to predict some of the content that should be assessed.

You should know

> The key messages from neuroscience about effective revision are that you remember what you think about and you should take regular breaks.
> Use flashcards, spaced repetition and loci to improve your ability to remember the specification content.
> There is a wide range of resources to improve your understanding of physics content. Explore them to find the resources that work for you.
> Use past papers, mark schemes and examiners' reports to get key insights into the style of thinking that will be required in your exams and evaluate your own level of understanding.
> Create mind maps, comparison tables, and model answers to help you see links between topics and gain a deeper understanding of the subject content.

Exam board focus

Learning objectives

> To provide an overview of the course content for each exam board
> To describe the assessments for each exam board
> To list typical question command words that assess higher-order skills
> To gain insight into the examining process
> To review good technique for multiple-choice questions
> To describe exam techniques to obtain a top grade

This chapter presents a very brief overview of the assessments for each exam board along with notes about how each board assesses the higher-order skills. Although the chapter is designed with a focus on individual exam boards, the final section on 'Advice for students' is relevant to all students.

The exam boards featured in this section are:

→ AQA
→ Edexcel
→ Eduqas
→ OCR
→ WJEC

AQA

Core content and assessment

Paper 1	Paper 2	Paper 3
Measurement and errors	Measurement and errors	Section A:
Particles and radiation	Thermal physics	All content
Waves	Fields and their consequences	Practical skills and data analysis
Mechanics and materials	Nuclear physics	Section B:
Electricity	Assumed knowledge from Paper 1	Option unit:
Periodic motion		• Astrophysics
		• Medical physics
		• Engineering physics
		• Turning points in physics
		• Electronics
2 hours	2 hours	2 hours
85 marks	85 marks	80 marks

Paper 1	Paper 2	Paper 3
34% of A-level	34% of A-level	32% of A-level
60 marks of short- and long-answer questions 25 marks of multiple choice	60 marks of short- and long-answer questions 25 marks of multiple choice	45 marks of short- and long-answer questions on practical experiments and data analysis 35 marks of short- and long-answer questions on an optional topic

All papers have roughly one third of the marks allocated to AO1 questions.

Paper 2 has an increased AO2 (53%) weighting and reduced AO3 (15%) mark allocation.

There are **12** required practical activities.

Assessment of higher-order skills

Questions assessing higher-order skills, such as application of knowledge or requiring analysis, evaluation or judgements to be made, typically have command words such as:

→ **Suggest** how [some experimental evidence can be used to confirm a theory]... (3)

→ **Deduce**, using calculations, whether *** is suitable for a certain application. (4)

→ **Compare**... (2)

→ **Derive** an expression to show that... (3)

Extended-response questions may contain these command words and phrases:

→ **Discuss** some of the problems associated with... (6)

→ **Compare** the [two diagrams]... (6)

→ **Discuss** how [a piece of equipment works or how a particular phenomenon works] (6)

→ **Discuss** which [from a selection of pieces of equipment is best suited for a particular application] (6)

Edexcel

Core content and assessment

Paper 1	Paper 2	Paper 3
Working as a physicist Mechanics Electric circuits Further mechanics Electric and magnetic fields Nuclear and particle physics	Working as a physicist Materials Waves and particle nature of light Thermodynamics Space Nuclear radiation Gravitational fields Oscillations	Questions on any topic in the specification Synoptic questions that may draw on two or more different topics Experimental methods and data analysis
1 hour 45 mins	1 hour 45 mins	2 hours 30 mins
90 marks	90 marks	120 marks
30% of A-level	30% of A-level	40% of A-level
Multiple-choice, short open, open-response, calculations and extended-writing questions	Multiple-choice, short open, open-response, calculations and extended-writing questions	Short open, open-response, calculations and extended-writing questions

There are **18** required practical activities.

Assessment of higher-order skills

Higher-level questions requiring some form of judgement are not always restricted to higher-mark questions. For example:

→ **Criticise** the student's investigation and conclusion. (5)

→ **Explain** the shape of the graph... (3)

→ **Criticise** the statement provided by the student. (3)

→ **Deduce** why one piece of apparatus may be more effective than another for a particular function. (2)

Extended-response questions may contain these command words and phrases:

→ **Explain** [this could be an experimental observation or a phenomenon from the specification...] (6)

→ Use the information to **discuss**... (6)

→ **Discuss** how... [a certain model can explain a set of observations] (6)

→ **Criticise** this extract. (6)

Eduqas

Core content and assessment

Paper 1	Paper 2	Paper 3
Newtonian physics	Electricity and the universe	Section A: Light and nuclei Section B: Choice of 1 out of 4 options: • Alternating current • Medical physics • The physics of sports • Energy and the environment
2 hours 15 mins	2 hours	2 hours 15 mins
100 marks	100 marks	120 marks
31.25% of A-level	31.25% of A-level	37.5% of A-level
Section A: 80 marks of short-answer and extended-answer questions with *some questions set in a practical context* Section B: 20 marks — one comprehension question	A mix of short-answer and extended-answer questions with *some questions set in a practical context.*	Section A: 100 marks of short-answer and extended-answer questions with *some questions set in a practical context.* Section B: 20 marks

There are **12** required practical activities.

Assessment of higher-order skills

You will be assessed on your ability to select, organise and communicate information and ideas coherently using appropriate scientific conventions and technical language across different assessment objectives. This means that higher-level questions requiring some form of judgement are not always restricted to higher-mark questions.

OCR

Core content and assessment

Paper 1 — Modelling physics	Paper 2 — Exploring physics	Paper 3 — Unified physics
Development of practical skills in physics Foundations of physics Forces and motion Newtonian world and astrophysics	Development of practical skills in physics Foundations of physics Electrons, waves and photons Particles and medical physics	Content from whole specification
2 hours 15 minutes	2 hours 15 minutes	1 hour 30 minutes
100 marks	100 marks	70 marks
37% of A-level	37% of A-level	26% of A-level
Section A: 15 marks of multiple-choice questions Section B: 85 marks of short-answer questions and extended-response questions	Section A: 15 marks of multiple-choice questions Section B: 85 marks of short-answer questions and extended-response questions	Short-answer and extended-response questions

In the final paper it should be noted that there is a reduced emphasis on AO1 (demonstrating knowledge and understanding) and AO2 (application of knowledge and understanding) assessment questions, and an increased emphasis on AO3 (analysis, interpretation and evaluation).

There are **12** required practical activities.

Assessment of higher-order skills

OCR has an emphasis on synoptic assessment because it believes it encourages students to see physics as a discipline rather than as separate topics. All three OCR papers therefore include synoptic assessment.

OCR includes level-of-response questions to assess your response in two strands:

→ Science content: your scientific knowledge and understanding and ability to apply this in unfamiliar situations.

→ Communication: your ability to communicate in a clear, coherent and logical way.

The examiner will read through your whole answer, before deciding the level that best fits your answer. The top mark band (level 3) means that you will be awarded 5 or 6 marks. The higher mark will be awarded if your answer is a good match to the main points, including the communication statement.

Level-of-response questions are indicated in question papers with an asterisk (*) after the question number.

Higher-level questions requiring some form of judgement are not always restricted to higher-mark questions. For example:

→ **Discuss** how the actual value compares with the value you calculated. (1)

→ **Suggest** the impact this may have on... (1)

→ **Explain** the variation [using information presented in graphical form] (3)

→ **Compare** this value with your value and **explain** why the values may differ. (4)

Extended-response questions may contain these command words and phrases:

→ **Describe** with the help of a labelled diagram [how a certain investigation could be carried out]... (6)

→ Use your knowledge and understanding of *** to **explain** these observations. (6)

→ **Explain** what is meant by ***. **Describe** how the graph can be used to determine the value of ***. (6)

→ **Evaluate** the information [when presented with a sample student analysis of some data] and the analysis of the data from the experiment... (6)

→ **Plan** an experiment to determine... (6)

Note that OCR sometimes uses multiple command verbs (two or three) within a single question, so make sure that you respond to each aspect of the question.

WJEC

Core content and assessment

Unit 1	Unit 2	Unit 3	Unit 4	Unit 5
Motion Energy Matter	Electricity Light	Oscillations Nuclei	Fields Optics	Practical examination: • Experimental task • Practical analysis task
1 hour 30 minutes	1 hour 30 minutes	2 hours 15 minutes	2 hours	Experimental task (1 hour 30 minutes) Practical analysis task (60 minutes)
80 marks	80 marks	100 marks	100 marks	50 marks
20% of A-level	20% of A-level	25% of A-level	25% of A-level	10% of A-level
Short-answer and extended-answer structured questions with some set in a practical context	Short-answer and extended-answer structured questions with some set in a practical context	Section A: 80 marks of short-answer and extended-answer questions with some set in a practical context Section B: 20 marks — one comprehension question	Section A: 80 marks of short-answer and extended-answer questions with some set in a practical context Section B: 20 marks — choice of one out of four options: • Alternating currents • Medical physics • The physics of sports • Energy and the environment	SPRING TERM OF 2ND YEAR OF STUDY Experimental task: 25 marks — you will be provided with a set of apparatus and an experimental problem Practical analysis task: 25 marks

There are **26** required practical activities (this is the most of any exam board).

WJEC is the only exam board which features 'direct assessment' of practical skills via an experimental task. In the experimental task you will be provided with a set of apparatus and an experimental problem.

All other examination boards apply 'indirect assessment' using questions in the written examinations. Note that all WJEC written papers may feature some indirect assessment of practical skills or data analysis.

Assessment of higher-order skills

Questions assessing higher-order skills, such as application of knowledge, or requiring analysis, evaluation or judgements to be made typically have command words such as:

→ **Stating any assumptions you make, show that**...   (3)

→ Use the graph to **compare**... (calculations are not required)   (4)

→ **Evaluate** the associated benefits and risks of...   (2)

→ **Draw** a diagram of a circuit you might use to investigate...   (2)

→ **Comment** on the quality and adequacy of the data obtained.   (2)

The quality of extended response (QER) question is explicitly identified on the front page of each examination. Inside the exam paper the question is also identified by **(6 QER)** for the number of allocated marks, rather than (6) after the question.

Typical command words for the extended-response questions are:

→ **Explain** how... [changing a certain factor affects this result, using information presented in the form of a diagram]   (6 QER)

→ **Explain** how a *** works.   (6 QER)

→ **Explain** how these data might lead scientists to conclude...

(6 QER)

Note that the extended-response question may have two separate ideas for you to explain. It might be more recall-based — 'explain a complex idea from the specification' — or it might require you to interpret experimental results.

Advice for students

The following section gives some general advice on the examining process and then lays out a series of top tips to ensure you maximise your performance in the exam.

General tips regarding the examination process

A little insight into how your papers are marked may help you write answers which communicate your ideas to the examiner more effectively.

Each one of your physics papers will be scanned (hence the need to *write in black ink* for better contrast in the scanned image) and stored electronically by your examination board. The questions are then broken into separate parts. This means that different examiners will mark different questions on your paper. In the few days after you have sat the paper, the chief examiner meets with other senior examiners to finalise the mark scheme. If a significant number of candidates have interpreted a question in an alternative way from that which was expected, then the alternative (if it contains correct physics) may be added to the mark scheme.

When an examiner logs into the marking software, they will choose a question to mark and will generally complete all of one section at a time. This process means they get to know the question they are marking very well, both in terms of what the mark scheme says, and what other candidates are writing. One effect of this is that if your definition isn't quite word perfect compared with the other candidates' responses, it will stand out and lose marks. Your writing needs to be legible. There are various settings that the examiner can use to make the image slightly easier to read, such as enlarging the scanned image or changing the contrast, but *if it cannot be read then it cannot score marks*.

Make sure you cross through any incorrect working. Failure to do so will result in you losing marks. It will look like you are not sure which answer or working is correct and are 'hedging your bets', leaving multiple possibilities on the page.

Multiple-choice technique

All exam boards use multiple-choice questions as part of their assessment. Multiple-choice questions allow examiners to test your knowledge and understanding across a broader range of the specification than would be possible only with extended-response questions. Here are some tips for good multiple-choice technique:

→ You should always answer the multiple-choice questions last. They are worth 1 mark each, but can easily use up more time than their mark allocation warrants. There is a danger that when aiming for the top grade you can spend too long answering the most difficult multiple-choice questions, but often this time is better spent on questions that are worth more marks.

→ Read the questions slowly and carefully, and be on the look-out for the 'not' statement — although it is written in a bold typeface, many candidates will still jump at the first correct answer they see.

→ Use the space available on the paper to do working out, such as writing, rearranging and solving equations. Too many candidates think they have to do multiple-choice questions in their head.

→ Be aware that some calculations will involve more than one idea, and may require the use of more than one equation.

→ Work through all the questions you can do easily first and come back to the questions that require more thought later.

→ Do not guess an answer unless you really have to. When time is extremely short near the end of the exam, and when all other approaches have failed, you can make an intelligent guess because marks are not deducted for incorrect answers.

→ You should practise as many multiple-choice past paper questions as possible (some are even reused from previous papers).

Ten top tips for exam day success

Chapter 5 described various revision strategies you can use to understand and remember specified content and skills throughout the course and particularly in the lead-up to exams. This section shares ten top tips for success on the day of the exam.

1 Get plenty of sleep the night before the exam.

An effective revision timetable can help you manage your time and reduce stress — see the Put yourself in control: organise your time section of Chapter 5 (p. 85).

2 Get there on time with all of your equipment.

The night before the exam, double check the time of the exam and prepare everything you need to take for the exam, including: your calculator, black pen (and spare), HB pencil, rubber, ruler, pencil sharpener and protractor. Make sure you have a reliable alarm clock and leave plenty of time to get to the exam hall.

3 Read the questions carefully.

Use the reading styles described in Chapter 2 to sift through the information presented in the question. Identify the command word(s), underlying physics, context and allocated marks.

4 Answer the specific question being asked.

Don't repeat the question. Don't just write what you know about the subject. Don't just repeat textbook explanations.

Do base your answer around a specific equation or principle and use the context and extra information in the question stem to apply your physics to the situation you are being asked about.

5 Use information provided in the question.

Use the information given in the question to inform your decision making, and make a clear judgement reflecting on, and comparing, any calculated values.

6 Use technical language whenever possible and avoid equivocal statements (such as 'the variable will change').

Candidates too often lose marks for not communicating their physics knowledge with enough precision. Key physics terminology may get used imprecisely or incorrectly, e.g accuracy, error, precision, resolution and uncertainty. Be clear about the meaning of key terms and use them correctly to ensure a succinct and accurate response.

7 Show your thought process clearly.

In quantitative questions this means setting out your working in a logical way so that your processes can be followed.

Drawing a labelled diagram can help you and the examiner see the situation clearly.

8 Be aware of any additional criteria that apply to your answer.

All physical quantities should have a value and units. Vector quantities must have a specified direction. Has the question asked you to write your final answer to an appropriate number of significant figures?

9 Use 'ramping' to gain momentum.

Each exam paper is typically designed to get more challenging as you progress through the questions. This process is called 'ramping'. This same ramping process occurs through each question too, so that the subparts in a question get progressively more difficult. This means that if you get stuck on the final part of a question, move on to the next question, which should be easier, so that you can maintain your momentum. You can return to the bits you missed out later on.

10 Keep an eye on the time.

You need to monitor your progress against the clock to ensure you don't run out of time. The different exam boards have slightly different time allocations for each mark, but they are all around one and a half minutes per mark. Ensure you have enough time to answer the longer questions and if you are struggling for time, use the concept of 'ramping' (see previous point) to target the earlier sub-parts of the questions, which are more likely to yield a quick mark.

You should know

> The main topics on for your exam board and how these are assessed
> Examples of typical question command words used by your exam board to assess higher-order skills
> The basics process of how your exam answers are marked
> The essential strategies for multiple-choice questions
> Effective strategies for success on the day of the exam